CH

D0712930

WORKING ON MARS

Voyages of Scientific Discovery with the
Mars Exploration Rovers

William J. Clancey

THE MIT PRESS
CAMBRIDGE, MASSACHUSETTS
LONDON, ENGLAND

MIT Press books may be purchased at special quantity
discounts for business or sales promotional use.
For information, please email special_sales@mitpress.
mit.edu or write to Special Sales Department,
The MIT Press, 55 Hayward Street, Cambridge,
MA 02142.

This book was set in Helvetica Neue Pro by
The MIT Press. Printed and bound in the United
States of America.

Library of Congress Cataloging-in-Publication Data
Clancey, William J.
Working on Mars : voyages of scientific discovery with the
Mars exploration rovers / William J. Clancey.
     p. cm.
 Includes bibliographical references and index.
 ISBN 978-0-262-01775-6 (hardcover : alk. paper)
1. Mars (Planet)—Exploration. 2. Mars (Planet)—Geology.
3. Mars Exploration Rover Mission (U.S.) I. Title.
QB641.C54 2012
629.43'543—dc23

                                     2011052650
10   9   8   7   6   5   4   3   2   1

For my parents, who provided an education so I might rove

Contents

Preface

The success of the Mars Exploration Rover (MER) missions begins to answer a question fundamental to space exploration: can people remotely conduct field science on another planet using a mobile, programmable laboratory? By virtue of its longevity—over eight years at this writing—and its complexity of organization and daily process, MER provides a wonderful and in many ways an unanticipated example for understanding and designing future planetary science missions. Through MER, a stable, sustainable enterprise has been developed and proven for reliably controlling two rovers with highly collaborative scientific and engineering teams. During these years, Spirit and Opportunity were driven a total of more than 40 km (25 miles) and took hundreds of thousands of photographs with tens of thousands of APXS readings, Mini-TES surveys, and microimages of more than a hundred abraded surfaces. We have come a long way from our first forays with robotic field science in the lunar Surveyor missions of the 1960s and joysticking Lunokhod 23 miles across the moon. MER is not the first investigation on a nonterrestrial surface, the first to use a programmed laboratory with multiple instruments and cameras, or the first to demonstrate programmed mobility. But it is the first to combine these capabilities and thus make field science on Mars possible, and it did so in a multiyear mission. The story of MER is the story of a unique human-robotic enterprise, with new scientific and engineering collaborations, and a prolonged journey together on a largely unknown planetary surface.

Origin of the Study of the Scientists' Practices and Experience

From early 2002 through 2005, I led a group of NASA Ames computer and social scientists who engaged in the MER project as "participant observers" of the science team—people who documented the mission in videos and field notes while providing advice about facilities, organization, and processes.[1] Our theme was human-centered computing, an approach for designing work systems that begins by understanding how people actually do their work. We studied conversations, meetings, tools, use of time, plans, schedules, charts, and software. During the mission itself, starting with Spirit's landing in early January 2004, our group was allowed to observe the scientists' meetings in mission conference rooms at the Jet Propulsion Laboratory (JPL) that were otherwise off-limits to visitors, including other NASA scientists and managers. The engineering work remained closed, but we had plenty to study. We videotaped and recorded many meetings through the ninety-day nominal mission. In particular, I was present for two weeks in February 2004 during a particularly exciting time when the outcrop in Eagle crater was being examined.

Stemming from this participation, the human-centered computing group wrote papers and gave presentations about the scientists' work, including a dissertation.[2] Having previously studied field geologists and biologists in Mars analog settings in the Arctic and Utah desert in ten expeditions, I was most interested in the scientific practices: how does working with a mobile, programmable laboratory change the nature of field science? How does being a member of the science team with a shared robotic intermediary change what it means to be a scientist?

From the first NASA/JPL press release, I questioned why MER was called a "robotic geologist" and presented my thoughts at a conference workshop, "The Human Implications of Human-Robotic Interaction."[3] Another participant, Sherry Turkle, Professor of the Social Studies of Science and Technology at MIT, subsequently commissioned a book chapter about MER and specifically on the topic of technology and identity. She posed the following questions:

- How does the technology affect the identity of those who use it?
- How does it come to make them see themselves, their work, and what makes them professionals differently?
- What other philosophical, social, or psychological issues does it raise that touches on areas of identity, such as relationship to their bodies, to the analog versus digital worlds?

This perspective turned out to be very fruitful. I decided to tell my originally intended story of "doing science with a rover" by emphasizing the scientists' firsthand experience in both individual and social dimensions. I complemented my recordings and knowledge of MER with interviews of six selected scientists in August 2006, focusing on these questions.[4]

In telling this broader sociotechnological story, I've been most taken by the personal experience of the scientists in identifying with the rover and identifying with the team—a blending of personal and group identity through a robotic system—as they move together on the surface of Mars. Unlike the earlier fixed-in-place Viking landers (1976–1977) and the more recent Phoenix lander (2008), or even the tentatively moving tiny Sojourner (1997), and unlike the many planetary orbiters and flyby spacecraft, Spirit and Opportunity are designed to *rove on the surface*. The scientists continuously choose where to go as their exploration unfolds. Thus the well-known space exploration metaphor of a "voyage of scientific discovery" is particularly apt for MER, suggesting the subtitle of the book.

Subsequently, in producing this book as an elaboration of the original article, I have interviewed two engineers and the Principal Investigator, Steve Squyres; added historical mission comparisons and MER statistics; included many photographs and illustrations of the scientific work; and analyzed MER's capabilities and limitations.

Explicating How the MER Exploration System Enabled Field Science on Mars

In this book, I suggest that we view MER as an "exploration system" for doing field science. With this framing, I ask, what aspects of the design of the rover and how it is operated account for the quality of the scientists' work? In explicating the daily process of commanding the rover and comparing it to other missions, we find a variety of themes—"being the rover," textbook style, personal and public concerns, aesthetic interests—that together reveal how the scientists actually manipulated the rover and why the programmable tools and analytic instruments worked so well for them. These complementary aspects are easily missed when we focus independently on the science team, the rovers' design, science operations, or the scientific results.

The MER exploration system constitutes a blended scientific practice: *exploring scientifically*, engaging in *field and laboratory* work by interweaving *human and robotic* operations, using a textbook method (*being systematic*) yet stopping to investigate interesting features (*being opportunistic*), and probing the landscape in a *kinesthetic, imagined experience*. Each part of these, such as using laboratory instruments in the field, poses a tension in the work; combined, they make this remote expedition an especially intriguing accomplishment.

Throughout the exposition, I have also sought to clear up some confusion about MER that I encountered in my research, including:

- Why "exploration" appears in the MER name
- How MER's "adaptive operations" differ from Viking, providing a qualitatively different way of doing planetary science
- How MER field science relates to early voyages of discovery and subsequent pioneering fieldwork by Alexander von Humboldt
- How the rover's and scientists' capabilities differ and relate to each other

I determined that to convey the nature of the MER scientific investigation, I needed to help the reader understand how MER's science operations, surface exploration, and robotics related to fieldwork on Earth and previous missions. This explanation begins by distinguishing between the operational process at JPL and the scientific fieldwork that was occurring on Mars. Indeed, arguing that not much had changed, one old-timer told me that MER mission operations occurred in the same facility built for Viking, Building 264 at JPL! In articulating these points, I learned more about Viking and indeed found many comparisons worth making with MER—but the similarities relate to the structure of the teams and processes on Earth, not the activities occurring on Mars through the rover.

Clarifying the nature of fieldwork and using a robotic laboratory on Mars—and how this is different from "mission operations" occurring on Earth—reveals the story of our first scientific field trip on another planet. The scientists' experience of "doing science with a rover" and what accounts for the quality of their work can be framed as elaborations of the corresponding topics listed previously:

- Scientific fieldwork in geology inherently involves exploring terrain by moving through it and opportunistically examining what you encounter.

- Daily science commanding (compared to Viking's two-week delay) enables the team to retain situational awareness, as they must react quickly and flexibly to operations in a changing landscape.
- Scientific work by the Forsters during Cook's second voyage inspired the notion of "scientific exploration"; it is not a coincidence or historically naïve that MER scientists named craters and major features after the ships of early explorers; their choices are clues about how they think about their work, and this shared historical framing facilitates their collaboration, as well as their anonymity in the mission.
- The quality of the mission's scientific work depends on aspects of MER's design that promote the *agency* of the scientists—their ongoing engagement in an inquiry that involves looking, manipulating materials, and moving on Mars.

Explicating the story of MER requires explaining the nature of field science on Earth (including, for example, undersea archeology using submersibles and teleoperated robots), the nature of exploration as a cognitive activity (as opposed to a social-political motive or period of history), and the factual differences of other surface missions (Viking, Surveyor, Sojourner, etc.) and orbital/flyby missions (e.g., Cassini). Focusing on what the scientists are doing on Mars, I investigate questions such as, "What is the relation of science and exploration?" and "How was controlling Viking's instruments different?" The answers reveal critical aspects of the MER exploration system that contributed to the expedition's success, and these are lessons we can build on for exploring the rest of the red planet, asteroids, and the moons of the gas giants that lie beyond.

To summarize the book's organization: chapter 1, "Scientists Working on Mars," introduces my initial observations of the scientists' work, raising paradoxes about field scientists working remotely and the presentations to the public that suggest the rover works independently. Chapter 2, "Mission Origin and Accomplishments," relates MER's design to other planetary missions and outlines the results and timeline of the MER campaigns. Chapter 3 "A New Kind of Field Science," explicates the nature of field science and how telerobotics changed the nature of scientific practices. Chapter 4, "A New Kind of Scientific Exploration System," explains how daily commanding, visualization tools, the rover's design, and team organization promote an experience of "virtual presence" that contributes significantly to the quality of the scientific work. Chapter 5, "The Mission Scientists," introduces the scientists and how the interviewees were selected, explaining how they blend personal expertise to form a team investigating Mars that is bolstered by their shared motivation to be explorers. Chapter 6, "Being the Rover: We're on Mars," describes how virtual presence is practically realized and experienced in the MER mission as "being the rover" through language, kinesthetic experience, and telerobotic tools. Chapter 7, "The Communal Scientist," explains how the "one instrument, one team" organization relates to both MER's physical design and the requirement for coherent fieldwork, emphasizing how inherent disciplinary conflicts were avoided or resolved. Chapter 8, "The Scientist Engineers," considers how the scientists and engineers collaborated in both complementing and adopting the scientists' interests. Chapter 9, "The Personal Scientist," then delves into nontechnical dimensions of the scientists' experience, both personal and public, with a particular focus on aesthetic experience. Chapter 10, "The Future of Planetary Surface Exploration," summarizes the main points and discusses MER's capabilities and limitations to guide the design of future scientific exploration systems. The epilogue revisits how, as the interviews with the scientists amply reveal, the motivation for such projects is always more than narrowly scientific and how a poetic presentation of our goals and accomplishments may be both necessary and valuable for realizing the vision of space exploration.

The book's subtitle encapsulates the experience of working on Mars: how the scientists *voyage* on the rover, traveling together with engineers; *exploring* the Gusev crater, behind Columbia Hills, and the Meridiani plains and craters; making *scientific discoveries* about what materials are present (or absent) and developing theories about how they formed.

By viewing MER from different perspectives, we come to understand it as a scientific exploration system that makes working on Mars possible: (1) an integrated combination of programmable sensors, effectors (e.g., an abrader), and instruments (e.g., for chemical analysis), (2) a team organization and work processes providing a capability for daily turnaround in receiving data and reprogramming these devices, and (3) semi-automated programming tools for collaboratively planning observations, including "virtual reality" tools for naming features and targeting instruments. The MER exploration system enabled a scientific process that was at once deliberative, opportunistic, and largely public. Other missions have had similar characteristics, but the combination of the MER exploration system, with its daily reprogramming, and the scientific work being accomplished on Mars—the first overland expedition on another planet—is brand new, in the words of Gentry Lee, who has participated on most planetary missions since Viking.

A key purpose of this book is to reveal, by plain speaking about what people are accomplishing with MER as conveyed through their personal experiences, that the practical relation of people and robotic systems is quite different than is typically implied in discussions about "robot explorers" or "partners" and the poetic description of MER as a "robotic geologist." These inspirational descriptions have a place, but we must understand this new technology and articulate how it has worked so well if we are to know how to use and improve it. Understanding MER requires a theoretical foundation that includes the concepts of virtual presence and agency, an understanding that requires parting the curtain to see what lies behind tales of "Spirit's struggles" and "Opportunity's discoveries." In addition to journalists, computer scientists and program managers can benefit, too: the rovers are tools, but the relation of the scientists to a programmable laboratory is far more complex than how a geologist relates to a hammer or, for that matter, how a principal investigator relates to a "payload" (the aerospace engineering name for a spacecraft instrument). This book combines perspectives from computer science, anthropology, philosophy, and cognitive science to build on the scientists' stories, yielding a better understanding of why the rovers worked so well and the lessons for designing future missions.

Orienting to the Scientific Fieldwork Occurring on Mars

So finally, to set the scene: this is a book about the scientific method of using a remote, programmable mobile laboratory to conduct fieldwork on Mars as revealed through the experiences of the scientists and by comparison to other missions and science projects. Other MER publications describe the development and testing of spacecraft instruments and sol-by-sol surprises,[5] scientific findings,[6] and the engineers' own realization of a dream.[7] Other mission histories might be written about how the MER project came about, such as institutional influences or the social milieu supporting the scientific study of Mars. As Ezell and Ezell said in their history of Viking, "In ignoring certain aspects or in describing others only briefly, we have not intended to slight other important aspects of the Viking effort. There are just too many stories and too many participants for them all to be included in this single volume."[8]

Although NASA mission histories typically describe engineering operations and provide personal perspectives, they have not focused on the scientific work in quite the same way attempted here. Certainly, previous missions have required new scientific practices, too, because of the need to perceive and act at a distance, time delay, new instruments and software, scientific collaborations across instruments and with engineers, multiple national organizations, and so on. Viking's mission operations especially showed the way.

But our more than eight years traveling on a rover has revealed another tale, an explication about field science and what it means to be moving through and exploring an alien surface, studying the aqueous, climatic, and geologic history of Mars. Understanding the differences between scientists and their tools and getting the language right are essential if we are to make good investment decisions in carrying forward this scientific exploration of Mars. So leaving behind the "robotic geologist" metaphor, you are invited to adopt a poetic image truer to how scientists related to these machines: climb into the rover with the MER team and voyage over the Columbia Hills and Meridiani's basaltic sands on the first field expeditions of another planet. As one scientist said, "We were all there, together, through a robot!"

Acknowledgments

I am indebted to the MER scientists and engineers who have shared their personal experiences in interviews and commented on the manuscript: Nathalie A. Cabrol, Michael H. Carr, David J. Des Marais, Jake Matijevic, James W. Rice, Michael H. Sims, Steve Squyres, Ashitey Trebi-Ollennu, and R. Aileen Yingst. Rice, Cabrol, and Sims also oriented my attention and understanding during the nominal mission at JPL in Pasadena, California, during January–February 2004.

The MER Human-Centered Computing ethnography team that I advised at NASA/ Ames included Charlotte Linde, Zara Mirmalek (University of California, San Diego), Chin Seah, Valerie Shalin (Wright State University), and Roxana Wales. Their observations, shared field notes, and our conversations played a crucial role in my understanding of MER operations. Charlotte Linde and Janet Vertesi generously commented on the original manuscript. Sherry Turkle encouraged writing with an ethnographic voice from the start; David Mindell asked similarly for more firsthand stories to begin the chapters, better fitting the personal style I wished to convey and leading to the serendipitous rediscovery of the story that begins chapter 9.

I benefited greatly from the reviews provided by the NASA Headquarters History Division under the guidance of Steve Garber. Conversations with Glenn Bugos, head of the NASA Ames History Office, helped me to recognize how a cognitive science perspective could go beyond the analysis of a traditional mission history. I deeply appreciate subsequent reviews provided by publishers, particularly for their attention to detail and strong advocacy for this work. David Mindell and an anonymous technical reviewer made perceptive and challenging suggestions regarding style, organization, and topics, which I have tried to take to heart. An anonymous JPL engineer who worked on the MER mission during both the system development and during nearly four years of the daily mission operations provided several factual corrections. John Callas and Andy Mishkin provided data about the JPL team and helped me understand the relations of staffing, budget, and operations. I also appreciate the friendly assistance of Dan Pappas in the NASA Ames Library and the excellent contributions by the MIT Press, including editor Kathy Caruso, copyeditor Nancy Kotary, and the book's designer Margarita Encomienda.

I gratefully acknowledge the continuing support and encouragement by Ken Ford and Jack Hansen at the Florida Institute for Human and Machine Cognition, my home institution, and by Mike Shafto at NASA Ames.

This work was funded in part by NASA's Computing, Communications, and Information Technology Program, Intelligent Systems subprogram, Constellation Program, and a grant from NASA's History Division. No royalties are received for this work. The US Government retains an exclusive, royalty-free license to use, reproduce, and have reproduced on behalf of the US Government the original manuscript on which this book was based. The views expressed herein are my own and do not represent the views of NASA or the US Government.

1

Scientists Working on Mars

Living with Spirit and Opportunity

It is 3:13 a.m. at Gusev crater on Mars, and the Mars Exploration Rover (MER) called *Spirit* is powered down for the night. The team of scientists who are "working Gusev" are living and working on Mars time, but, with some luck, they are fast asleep in their Pasadena, California, apartments. The MER is a remotely operated vehicle, and it is not the only one exploring Mars at this time. On nearly the opposite side of the planet, at Meridiani Planum, it is 3:12 p.m. local Mars time and another MER called *Opportunity* is busy carrying out its programmed plan of photography, mineralogic analysis, and driving to another site. Meanwhile, the thirty-seven scientists "working Meridiani" are ensconced in a fifth-floor meeting room at the Jet Propulsion Laboratory (JPL) in Pasadena, planning what Opportunity will do tomorrow.

Although it is 2:36 p.m. on a bright February afternoon in Southern California, the MER scientists sit in a dark room with heavy black shades covering huge windows—making it easier to see the screens, but also helping them mentally project themselves into the local Mars time zone of their rover. They are fortunate that the clocks at Pasadena and Meridiani are nearly synchronized today; as both Earth and Mars rotate, these lands are pointing to the sun together. But each Mars second is slightly longer than an Earth second,[1] so tomorrow, to passersby in Pasadena and the JPL gate officer, the scientists will appear to report to work forty minutes later than the day before. And in a few weeks, the Meridiani team will be driving to work at midnight, then 12:40 a.m. the next day, and so on. By then, the Spirit team will be fighting lunchtime traffic on their commute to work. With the heavy shades on the windows, the MER scientists block out their Earth-bound existences and place themselves with their rovers on Mars. Indeed, in describing a rover's activity, they do not even mention days, which refer to the rotation of the Earth. Rather, they count time by the number of *sols*—Mars rotations—since the landing at their particular site. Tosol (this solar day on Mars) is sol 25 at Meridiani (M25)—twenty-five Mars sols have passed since Opportunity landed.

The MER scientists will be living on local Mars time for three months, the "nominal" mission. Each rover usually completes its daily program and "sleeps" about midafternoon local Mars solar time, and the scientists working that mission appear for work at JPL shortly thereafter. Schedules depend on personal preference and the relation of Mars to Pasadena time, but typically scientists have dinner and then sleep after work as a third shift of engineers prepares the uplink; then the scientists awaken about the time that their rover begins work on the next sol. For six months after the nominal mission, the science and engineering teams were no longer colocated; instead, the scientists participated by phone from their home institutions.[2] Operations also shifted after three months to allow weekends off, with rover programs planned several sols in advance. In the years that followed, the team was fully distributed (including more than eight groups), with the planning becoming more streamlined and conforming to "skeletal" sequences or templates.[3]

CHAPTER 1

On sol 25, I am working Meridiani. Working with a rover during the nominal mission involves a bewildering complex of meetings, organizations, and schedules. The first meeting of the sol, the science context meeting, includes short lectures by half a dozen scientists. Using noisy projectors (which require them to speak with wireless hand microphones to be heard in the cavernous room), they display dozens of colorful photographs, charts, and bulleted plans with titles like "Locations and Things to Do for Mineralogy, or What Kinds of Targets We Need" and "Geologists Stop for…." Organized into four science theme groups (STGs) and a long-term planning (LTP) group, they sit in clusters around heavy wooden tables with large computer displays like black boxes requiring them to stand, sit on their knees, or scoot around on their rolling chairs to see each other. Later in the sol, at the science assessment meeting, the Science Operations Working Group (SOWG) chair polls the STGs clockwise for "their sequences"[4]—Mineralogy and Geochemistry, Soils and Rocks, Geology, Atmosphere. Usually standing, each STG lead offers initial suggestions about tomorrow's work on Mars. These ideas will become plans that STG leads will merge before and during the SOWG meeting with the engineers after dinner. The agreed-upon plan will be transformed by the engineers into an "upload sequence," the computer code for the rover's next-day work on Mars. The scientists' end-of-sol meeting will then review the rover's progress and tomorrow's plan and reflect on where the group is headed (figure 1.1/plate 1).

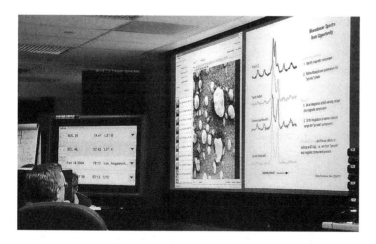

FIGURE 1.1/PLATE 1
Sol 25 in Meridiani science assessment room end-of-sol meeting (February 18, 2004). *Note:* Unless otherwise noted, photographs and graphic reproductions are by the author.

Today, on sol 25—a month into Opportunity's nominal mission—the SOWG chair, Steve Squyres, is giving a short lecture to the scientists about using the rock abrasion tool (RAT). The RAT is by design analogous to a geologist's hammer, though it works more like a very wide drill, scraping a circle into a hard surface (figure 1.2/plate 2): "As we think about how we are going to approach this outcrop, our thinking and our discussion should be very much based on hypothesis testing.... What's the scientific hypothesis that you're trying to test and why are we using the RAT or talking about using the RAT in the first place?"[5] The chair continued to explain how choosing areas with different colors might reveal different underlying compositions: "We should find the purplest stuff and greenest stuff that we can and RAT the hell out of both them and see if they're different underneath them" (where the colors refer to those used in graphic visualizations of spectral data, as in figures 1.1 and 6.10). After further explaining what the RAT is designed to do, he leads the team to anticipate what they might see: "I think there is a good use of a RAT on a place where you think to see cross-bedding because ... that makes some specific predictions about the grain sizes of the stuff that's involved." He concludes, "So all I'm saying is as you think about this, don't say let's RAT here to reveal this, and let's RAT here to reveal that, let's talk it through in terms of the specific scientific hypotheses that we're trying to test."[6]

Having studied geologists working in Mars analog landscapes in the Arctic and Utah, I have a stark realization: we aren't on Earth any more. Why is it necessary to tell these scientists—who are experienced in field exploration, who have been competitively selected to be members of the MER Science Team, and many of whom have worked on Mars orbiter missions or with the short-lived Sojourner rover in 1997—how to do science?

No field geologist on Earth explores in this way—articulating hypotheses before every hammer swing, before each rock is opened. This looks like textbook science: form a hypothesis or two, design an experiment, test, and repeat. Like professionals asked to tell you what they do, the work appears overrationalized.[7] Is this correct? Does working with a rover actually require changing the practice of field science?[8]

The Paradox of Exploring with a Rover
Considering the large team assembled in the science meeting, how could field scientists tolerate this communal style of investigation? These are people who have often told me that they chose their careers because they like to be outdoors. How does it change their sense of their profession and their personal life to work for three months in a room with heavy black shades, participating in an apparently endless series of meetings, seven sols a week, living out of an apartment, away from their family? This is what working with a rover on Mars entailed in 2004 and as a somewhat less public but still communal activity continued in teleconferences for years thereafter.

Geologists and other planetary scientists are people who crave climbing hills and hanging onto craggy outcrops, often traveling kilometers in a single day. How does it feel

FIGURE 1.2/PLATE 2
One chosen placement for the rock abrasion tool (M27). "The rover visualization team from NASA
Ames ... initiated the graphics by putting two panoramic camera images of the 'El Capitan'
area into their three-dimensional model. The rock abrasion tool team from Honeybee Robotics
then used the visualization tool [Viz] to help target and orient their instrument on the safest
and most scientifically interesting locations." Description from JPL MER Mission, "Plotting and
Scheming," http://www.jpl.nasa.gov/missions/mer/images.cfm?id=1169 (accessed February 20,
2004). For information about Viz, see Laurence Edwards, Judd Bowman, Charles Kunz, David
Lees, and Michael Sims, "Photo-realistic Terrain Modeling and Visualization for Mars Exploration
Rover Science Operations," *Proceedings of IEEE SMC 2* (Hawaii, October 10–12, 2005): 1389–
1395. *Image credit:* NASA/JPL/Cornell/Ames/Honeybee Robotics. *Source:* JPL, "Photojournal,"
http://photojournal.jpl.nasa.gov/catalog/PIA05338 (accessed February 20, 2004).

to explore while virtually sitting on a rover that requires more than three or four years to go 10 km? The pace is slow and painstaking; how does the work keep their attention? How can they keep track of what they are doing from day to day? How do you remember what's behind you and whether you need to move the rover for your arm to reach a target or to see a more distant feature?

These are people who like to put their fingers in sand and scratch, smell, and even taste rocks. How is it possible to know a planet's geology from digital photographs and charts on a computer screen? These are people who often work alone or with close colleagues, relishing the intimacy of expeditions and their personal investigations. What is it like being shackled at the hip, as it were, traveling with specialists in different disciplines, exploring and surveying new terrain, systematically gathering data for future scientists to analyze? How can they tolerate depending on other people: the engineers, who serve as the scientists' agents in commanding every motion and measurement on Mars?

Clearly, working with a rover on Mars changes many of the essential aspects of field scientists' exploration on Earth: direct presence, pace of work, autonomy, and privacy. But paradoxically, the scientists feel as if they are on Mars doing field science. The daily commanding cycle and visualization tools for targeting observations provide an experience of virtual presence, of *agency* in manipulating tools, scuffing the surface, shifting views, and traversing great distances over hills and plains. One scientist said it is as if he has "two boots on the ground."

Through the combined system of tools, programming, roles, and schedule, MER has changed the nature of planetary field science, enabling the first scientific exploration of a broad terrain on another planet. To understand how doing science in this manner is possible, the scientists' experience, and lessons for future missions, we need to consider in some detail the nature of field science, scientific exploration, telerobotics, and the one-sol turnaround in receiving data and uploading plans.

By late summer of 2006, the two MER rovers had exceeded their nominal mission milestones by a factor of ten: worked for more than 900 sols, driven nearly 16 km (almost 10 miles[9]), taken tens of thousands of photographs, and so on. By mid-2009, the number of sols had doubled, each rover had taken more than 100,000 images, and Opportunity itself had gone farther than 16 km. By mid-2011, with Spirit now quiescent, the number of Opportunity images was nearing 150,000, and it had traveled over 30 km. In opening up new landscapes in an enticing but dangerous environment, the MER expeditions remind us in some respects of the early scientific exploration of Earth. The difference of course is that the scientists are exploring Mars, but nobody has traveled farther than Pasadena.

The Paradox of the Robotic Geologist

Beyond the questions raised about doing fieldwork without being physically present, a still stranger curiosity developed during the MER mission: the rovers became the heroes of the story. The official JPL website announced when the first rover landed: "Spirit Lands on Mars and Sends Postcards: A Traveling Robotic Geologist from NASA Has Landed on Mars and Returned Stunning Images of the Area around Its Landing Site in Gusev Crater."[10] Posted mission reports regularly have ascribed machine initiative to remotely controlled actions: "Spirit collected additional imagery of the right front wheel."[11] When something goes wrong or a problem is solved, attention shifts to the people behind the scenes: "The operations team has successfully commanded Spirit to drive using only 5 wheels."[12] Making the rover into a person—that is, anthropomorphizing it—began publicly in the first press release:

> July 28, 2000—In 2003, NASA plans to launch a *relative* of the now-famous 1997 Mars Pathfinder rover. Using drop, bounce, and roll technology, this larger *cousin* is expected to reach the surface of the Red Planet in January, 2004.... This new *robotic explorer* will be able to trek up to 110 yards (100 meters) across the surface each Martian day.... "This mission will give us the first ever *robot field geologist* on Mars ...," said Scott Hubbard, Mars Program Director at NASA Headquarters.[13]

Besides casting the machinery as part of a biological family of relatives and cousins, this statement described the rover with phrases that appeared throughout the mission: the robot is "an explorer" and "a field geologist."

How could a rover having no ability to plan, perceive, interpret, or theorize be described as a geologist? Unlike a surrogate that replaces a person (the robots of science fiction), the rover is only a mechanism that is "acted through," an extended embodiment of the human eyes and hands of the people who control its actions from Earth. Yet press reports might lead one to just the opposite impression: if NASA has already sent robotic geologists to Mars, why do we need to send people into space?

To understand why critically examining the "robotic geologist" mindset is essential for understanding the MER mission, consider Matt Golombek's 2007 article in *The Planetary Report*, "Spirit and Opportunity—Martian Geologists."[14] Golombek, a lead MER scientist and the Pathfinder project scientist, provides a superbly readable technical summary of geology findings, concisely explaining the morphologies and combination of processes that formed, mixed, and altered rock and sand materials near the MER landing sites. He describes the rovers: "These mobile robots have been our field geologists—traversing … and examining the surroundings.... During their journeys, the hard-working rovers have returned compelling evidence."[15] Amid the clear technical presentation, he mixes poetic attributions about the rover's inquiry (italicized) with the scientists' inquiry (underlined): "After exploring Endurance crater, *Opportunity drove* south to *investigate* the heat shield it used during landing. Next to the heat shield, <u>we noticed</u> the only rock seen for kilometers on the plains. *Opportunity's investigation* of this rock revealed it as a nickel iron meteorite, a very exciting finding, as it was <u>our first discovery</u> of a meteorite on another planet. (Since then, *Spirit has discovered* two others.)"[16] The text continues to describe "the rovers' explorations." Golombek says, "Opportunity encountered" cross beds, "found desiccation cracks," "is currently investigating the stratigraphy," and "will study for quite some time." The team adopts this same style in *Science* articles.

In the robotic geologist metaphor, perhaps the most intriguing irony and puzzle of the MER mission, we find a deeper truth about the difficulty of communal and virtual science, a truth about the human ability for projection and imagination that in many respects made working with the rover possible and the invisibility of being a member of the science team tolerable. Indeed, we find that for some settings and purposes, different points of view are adopted—the rover as "we" and "my partner." A key part of my inquiry is to understand why the scientists and engineers adopt different ways of talking, how these are actually useful perspectives for operating the rovers, and what is being obscured by the difficulty of grasping and describing a new kind of scientific exploration system.

Deferring credit to the "hard-working" rover, the scientists have adopted the academic style of a technical publication: scientists focus on and write about the phenomena they study, not themselves and their travails. But the resulting presentation is like the Wizard of Oz's hypnotically imposing image: all eyes are on the "intrepid explorer"[17] onstage, with no attention paid to the team behind the curtain. This heroic point of view makes an engaging story for nonscientists while conveying many facts.

Yet, in the end, what do we know about the scientific investigation? Regarding how Viking was presented in the 1970s, "many of the young people [Hal Masursky] had talked with had thought NASA's unmanned space projects were controlled by one great computer with no human beings involved."[18] Imagine the questions a curious teenager might ask on reading the article from *The Planetary Report*: are the rovers merely wandering about and encountering rocks randomly? Have they been programmed to look for something in particular? Do the scientists have a say in how long Spirit will be studying Home Plate? Metaphors are useful, but not to the point that they obscure the essential nature of the mission—that scientists on Earth are conducting fieldwork over a large and varied terrain of another planet using a programmable robotic laboratory.

Talk about new technology often uses metaphors that attempt to express something new in familiar words, such as calling a car a "horseless carriage." The word "automobile" was born in the paradox of a vehicle that didn't require a horse (or person) to pull it—a self-propelled mobile vehicle. With today's robotic technology, we might expect an "automobile" to not require a driver, either (and after one hundred years of development, this might soon happen). Referring to the rovers as "geologists" pretends not only that they are the drivers, but also that they have places they'd like to go and their own strategies for roving Mars. Someday that may be possible, but not now.

At issue is the nature of *agency*: who is using the tools? Who is interpreting the data and claiming discoveries? The MER scientists and journalists speak poetically to make their story understandable but also because this technologically enabled practice, with its complex combination of tools and operations, changes the position and role of people in the physical work. Making the robot the actor avoids all the details, focusing on the exciting discoveries. But then we don't learn what is easy and what is hard, and why the rover worked so well—the craft and art of doing fieldwork remotely is untold. My intent in this book is thus to provide an appropriate scientific framework for thinking about the scientists' experience as actors on Mars to help us articulate the synergies of people, technology, and process and what can be improved.

Learning from MER by Analyzing the Scientists' Experience

To see how learning from MER requires a proper analysis of agency in remotely conducted field science, why this analysis is challenging, and how we might proceed, consider how Squyres expressed a potentially crucial limitation of MER: "Our rovers have discovered some wonderful things, but not the kinds of things humans could."[19] Clear speaking about the rovers begins by focusing on the scientists' inquiry and restating Squyres's remark, perhaps as follows: "Using robotic laboratories, remotely controlled from Earth, scientists have discovered some wonderful things, but not the kinds of things people could discover if they were actually on Mars." This way of summarizing the fieldwork raises a complex scientific question that psychologists, philosophers, and computer scientists alike have stumbled over the past fifty years: "What can people do that robots cannot do?" But that is not the scientific question at hand for mission planners of the next few decades, which is, "What kinds of things could people on Earth investigate better using a different robotic laboratory on Mars?" We need to ask the science team, "What did you want to do at particular times and places with certain materials that MER did not allow?" Or, if you wish, "What would scientists do on Mars that cannot be accomplished by controlling robotic laboratories from Earth?"

Before we can evaluate the answers to these questions, it is evident from the MER scientists' natural focus on scientific results rather than their work that a brief tutorial about scientific inquiry, and fieldwork specifically, might be useful. My approach is to start the discourse from scratch and explicate the cognitive work of exploring and explanatory modeling that occurs in surveying and traversing a new topography. From that basis, we can separate out the entertaining rhetoric about Spirit and Opportunity's "hard-working investigation" as "intrepid explorers," which conveys the scientists' emotional experience both in the small, in personally laboring through a tool they designed and directed, and in MER's broader sociohistorical meaning—to formulate a philosophically astute description about the scientific method as it is applied and developed in this mission. We can do this by focusing on the scientists' experience in doing fieldwork on Mars using a robotic laboratory.

Although each of the characteristics of using a rover on Mars for scientific inquiry— virtual presence, communal work, deliberative science, programmed instruments and roving, and others—is not new to MER, the synergistic combination of instruments, the sophistication of the programming and visualization tools, the duration of the mission and distances traversed, the collaboration between scientists and engineers, and the daily commanding cycle have in combination created a new system for scientific exploration of a remote landscape. Accordingly, the experience of being on a mission has also changed. By considering all aspects of "what it's like to use a rover on Mars," from the logistic through the aesthetic and deeply personal, we can get a well-rounded picture of MER's instrument integration, team structure, and operations process.

Accordingly, to cast a wide net, my conversations with the MER scientists cover a range of concerns and enthusiasms relating to the scientists' professional identities and their virtual presence on Mars, represented by chapter 5, "The Mission Scientists"; chapter 6, "Being the Rover: We're on Mars";[20] chapter 7, "The Communal Scientist"; chapter 8, "The Scientist Engineers"; and chapter 9, "The Personal Scientist." We find a surprising commonality of experience across the team despite differences in age, mission experience, and specialization. These scientists are highly adapted for their roles, but all must reconcile the mission's scientific, technical, and physical demands in what we might call the social construction of scientific productivity. They speak up for their disciplinary interests, analyze data on the side, and make instructive presentations to each other and the public. Thus they establish their identities as valued, authentic members of the team, as constrained by the limits and affordances of the MER technology.

One of the repeated findings, as Squyres's remark intimates, is that although the MER scientists are elated by the feeling of "being on Mars," they are frustrated by their tools. To evaluate these complaints, we need to understand their expectations by comparing their field and laboratory experience on Earth to working through a rover. We need to separate expectations from what tools they actually need (both on Earth and on Mars) to carry out this particular scientific inquiry around Gusev and Meridiani. Tool requirements are influenced by cognitive and social motives relating to the rhythm of work, personal interests, and standards of the scientific community—any of which could be frustrated by the MER exploration system. After understanding why the scientists are elated and why they are complaining, we can better evaluate how MER might be improved and indeed whether having people on Mars (with robotic laboratories at hand) is scientifically necessary or motivated only by other aesthetic-psychological or social-political desires.

The MER scientists have forged a team that explored Mars as a group, coaxing their rovers through treacherous sand and into dangerously steep rocky craters and teasing out the geologic and climatic story of another planet. Our challenge in learning from this "voyage of scientific discovery" is to put it in the historical perspective of planetary field science and to recognize how the system of technology, cognitive processes, and social participation fits together and then to see what we can conclude about strategies for exploring an entire planet.

2

Mission Origin and Accomplishments

Kneeling alongside the Mars Exploration Rover, you experience its human proportions—you can reach down and compare your arm to its arm for deploying instruments; you can stretch your arms to show the width of its wing-like solar panels; and standing, you are face to face with the panorama camera's two eyes, which are about 5 feet high. The invention of this robotic laboratory is complex historically, technically, and politically. Here I establish some basic facts about its design and place in the study of Mars, so that we can appreciate why the scientists are so excited to participate on this mission.

Putting the rover in perspective means understanding especially how it enables planetary field science. We begin by reviewing the history of Mars missions and the scientific questions that have been pursued. Focusing on MER, we consider how the rover concept developed, what has been accomplished in the various campaigns, and a timeline of events.

Planetary Surface Missions

Percival Lowell's misinterpretation of telescopic views of Mars as canals set off the twentieth-century study of Mars as a provocative quest whose essential nature continues today. The question about possible life on Mars has of course shifted dramatically from imagining unseen beings who created canals to imagining unseen microorganisms—perhaps buried in yet-unseen aquifers or whose fossils and effects reside in contentiously interpreted meteorites. Driven by an enticingly familiar desert landscape, the people of Earth have sent more than forty missions to Mars in the past forty-five years. Better imagery from orbit suggesting water erosion, combined in 1997 with the dramatic public interest in Sojourner, the first martian rover, revitalized NASA's commitment: "The Mars Program was created in the late 90s as enthusiasm for the search for life beyond the Earth was at a historic high … driven by the discovery of planets around other stars … Hubble images … and the Martian meteorite ALH84001."[1] For the scientists proposing new missions, the investigation needed to be done on the surface; as Squyres later reported, "Science I was interested in on Mars was very difficult to do from orbit…. From my background in geology … I liked the idea of reading the story that the rocks had to tell us."[2]

With an overall interest in understanding whether life could have existed on Mars, the scientific objective of the MER mission proposal was to study the "aqueous, climatic, and geologic history of sites … where evidence of possible prebiotic processes might have been preserved."[3] Squyres later recounted how "the most significant and exciting discoveries were clearly the ones related to water."[4] The strategy to "follow the water" was common to the geologists and astrobiologists, based on the features observed from orbit, Viking's finding that martian soil was reactive in water, and the ambiguity of the Allan Hills meteorite. ("Soil" refers to loose, unconsolidated materials; no organic, biological relation is implied.[5])

Squyres has explicated in detail the complicated, decade-long development of MER's design in a series of proposals. He emphasizes the evolving concepts of the instrument suite and changing risk-reward trade-offs following from success of Pathfinder/Sojourner and the lessons of costly failures.[6] Briefly, the context for defining and developing MER was both technological (in view of the failed missions of the 1990s) and scientific, as a planned Mars program for a series of missions announced in October 2000 included reconnaissance, increasingly sophisticated robotic laboratories, and an eventual sample return mission.[7] In both respects, MER could be viewed as a pivotal mission—to land successfully with a much heavier and more sophisticated "payload" than Pathfinder/Sojourner, and to pave the way for wide-ranging and long-term fieldwork on the martian surface. These stakes increased anticipation (and anxiety) and made the successes more emotionally overwhelming.

To understand the origins and accomplishments of MER as a scientific exploration system, as a means of studying a terrain by traveling through it, we begin by relating it to other planetary surface missions (table 2.1). Apollo 17 is included for comparison of the duration of surface operations and distance traveled. The astronauts on Apollo were indeed conducting lunar field science. Apollo 17's traverses for example totaled about 36 kilometers. But the surface investigations lasted less than a day for each mission, so Apollo extravehicular activities (EVAs) did not constitute an "expedition" in the sense familiar to Earth field scientists. Their paths and activities were planned and trained for down to the minute, though many adaptations occurred in practice.

The ability to control an electromechanical system on another planetary surface during a mission varies greatly from fixed, stored programs (no control during the mission) to direct control (by being in the same location), remote control (transmitting control signals), and batch processing (regularly transmitting programs). In this section, we consider the general historical relations of the missions; see chapter 4, "A New Kind of Scientific Exploration System," for an elaborated technical comparison of Viking, Sojourner, and MER. Teleoperation of Lunokhod and Surveyor is described in the section "Coordinating Telerobotic Inquiry: Visualization and Autonomy" in chapter 6.

The Soviet Lunokhod lunar rovers were the first remote-controlled mobile laboratories on another planetary surface. Viewed as a response to Apollo, the Soviet's concept of operations is historically interesting, but their approach is also relevant to the question of what scientific work can be accomplished by scientists working from Earth:

> Having lost the race to the Moon, the Russians set out to prove their robot technology could match the science any astronaut could do on the lunar surface. Luna 16 touched down on Mare Fecunditatis (the Sea of Fertility) September 20, 1970, about 60 hours after local sunset—the first lunar night-landing. Just 26 hours after arriving, Luna 16 had drilled into the lunar soil, retrieved a 13.7-inch-long (35 centimeters), 3.6-ounce (101 grams) geological sample, stowed it in a return capsule, and launched it toward Earth. The sample arrived September 24, floating on parachutes to a gentle touchdown

TABLE 2.1

Comparison of successful planetary surface missions.* Five Surveyor missions are grouped; duration here is the cumulative period of science operations. Seven Venera missions were successful, surviving one to two hours. Spacecraft and instruments are controlled by a preloaded fixed program or by sending commands for immediate operation (remote control) or delayed operation (batch programming). Each Viking lander was intensively programmed only during the first few months. "Commanding turnaround" is elapsed Earth time from preparation of a science plan to uplink of batch** commands to the telerobot. For comparison, Apollo 17 distance and duration refers to the lunar rover driven by astronauts.

MISSION	PLANETARY BODY	LANDING DATE	OPERATIONS DURATION	DISTANCE TRAVELED (KM)	CONTROL METHOD	COMMANDING TURNAROUND (DAYS)
SURVEYOR (5)	MOON	1966–1968	~4.5 MO	—	REMOTE	—
LUNOKHOD 1	MOON	NOV. 17, 1970	322 D	10.5	REMOTE	—
APOLLO 17	MOON	DEC. 1972	<1 D	35.9	DIRECT	—
LUNOKHOD 2	MOON	JAN. 15, 1973	120 D	37	REMOTE	—
VENERA 7–14	VENUS	1970–1981	~1 HR	—	FIXED	—
VIKING 1	MARS	JULY 20, 1976	>6 YR	—	BATCH	16–20
VIKING 2	MARS	SEPT. 3, 1976	>3 YR	—	BATCH	16–20
SOJOURNER	MARS	JULY 4, 1997	86 D	<0.1	BATCH	1
SPIRIT	MARS	JAN. 4, 2004	>6 YR	7.7	BATCH	1
OPPORTUNITY	MARS	JAN. 25, 2004	8+ YR	> 34	BATCH	1
HUYGENS	TITAN	JAN. 14, 2005	~1.5 HR	—	FIXED	—
PHOENIX	MARS	MAY 25, 2008	156 D	—	BATCH	2

*For mission data, see NASA Space Science Data Center, http://nssdc.gsfc.nasa.gov/nmc/SpacecraftQuery.jsp (accessed July 28, 2009).

**Before the advent of personal computers, using "mainframe" computers in the 1960s and 1970s, programmers wrote computer code in decks of Hollerith cards or on paper (and sometimes magnetic) tape, which were collected by computer operators into "batches" that were input to the computer and run in sequence. Output, usually in the form of line printer listings and cards or magnetic tapes, was then individually packaged and placed in bins for the programmers to pick up, often hours or a day later. The process of collecting programs into a batch and the turnaround time parallels at a high level the process of operating the MER computers. Of course, in almost every other aspect, operating a physical system on another planet—with multiple devices being commanded in parallel and sharing a common memory, power, and communication systems—is greatly different from mainframe programming.

in Kazakhstan. In the shadow of the successful manned U.S. Apollo 11 and 12 Moon missions, Soviet scientists had collected their first lunar sample—and they did it with a robot.[8]

Lunokhod 1 ("moonwalker" in Russian), launched on Luna 17, was the first mobile science laboratory on another planetary surface. Lunokhod 1 (1970) bears some comparison to MER in duration (322 days), distance traveled (more than 10 km), and instrumentation as a laboratory (multiple cameras, a spectrometer, and penetrometer). The Lunokhod 2 mission (1973) was shorter (4 months) but went much further (37 km) over more difficult terrain and included more instruments (soil mechanics tester, solar X-ray experiment, astrophotometer, magnetometer, radiometer, and photodetector). The Lunokhod rovers were teleoperated in real time, including a control stick ("joystick") for driving, rather than being sent daily programs.[9]

No further mobile science laboratories were sent into space until Sojourner, twenty-four years later. Although the scientific work appears superficially similar to Lunokhod—a camera and an alpha-particle X-ray spectrometer (APXS) on a rover—the technology had advanced considerably through computerized miniaturization and more sophisticated programming. The Soviets had used off-the-shelf instruments, placing them in pressurized vessels. Delivering the Lunokhods' average 800 kg (1763 lb) bulk (comparable to the Mars Science Laboratory) to Mars requires a powerful and costly launch, entry, and landing system. The useful scientific work we can do on Mars has been greatly enhanced by advances in packaging, automation, and instruments. Viking and Sojourner were thus the stepping-stones to MER.

Inventing a Mobile Robotic Laboratory

With Viking, we learned how to test materials on Mars using a programmed laboratory. Gerald Soffen commented in 1978, "How remarkable! We are performing chemical and biological experiments as though in our own laboratories. Taking pictures at will, listening for seismic shocks and making measurements of the atmosphere and surface."[10] Sojourner demonstrated the technical practicality and advantages of placing the instruments on a rover enabling rudimentary field science. Jake Matijevic, the engineering team chief for MER operations,[11] who focused on systems engineering throughout the mission, explains how the MER concept developed:

> It may be best to think in terms of the history of rovers on Mars. MER was an obvious extension of the results of the Viking mission. Looking at the two sites where the Viking landers landed, the scientists could see things they really wanted to do, that were just not possible given the fact that you were on a lander. Growing out of that scientific interest were the first ideas about how one could try to put a moving platform on the surface of Mars. In the late 70s and 80s the technology really wasn't there for trying to build a device that could work like this on the planets.[12]

By the late 1990s, microcomputers and other miniaturization of electronics for instruments made it possible to have a "payload with functions of roving on the surface," leading to the $25 million Sojourner rover. Compared to Lunokhod, Sojourner was much smaller and lighter (10.5 kg, 23 lb) and used less power. With its far more powerful computers, it could make calculations quickly enough to move on the surface and approach targets without direct human supervision. MER's daily operations schedule of analysis, planning, and programming—including living on Mars time—was prototyped during the Sojourner mission in 1997. As Sojourner's rover manager, Matijevic formalized the daily scenario of activities for the rover driver and sequence planner to translate into a program of commands.[13] This process was a transformative experience for JPL engineers and inspired the scientists to propose new rover missions that would be far more ambitious:

> Sojourner was intended as a demonstration. Given its limited objectives as a payload on the Mars Pathfinder lander, it sparked interest in the science community to expand the applications to a more complete instrument set, with the kinds of missions that we had only dreamed about from the Viking days—exploration missions, sampling missions, geological missions. It was a paradigm shift, when you think in terms of scientists who have done missions, built instruments, and operated them here [at JPL], for many, many years.

Sojourner stayed within 15 m of the Pathfinder lander (although 500 m was theoretically possible), was driven about 90 m, lasted less than three months, and was directed mostly by the engineering team.[14] In terms of exploration and field science, MER is in a different league: segments of Spirit's traverse to Home Plate, such as the campaign to the Columbia Hills (figure 2.1/plate 3), had their own coherent organization governed by strategies for systematic investigation adapted for each topography (e.g., outcrops in hills versus basalt stretches), and each campaign was longer than Sojourner's entire operation. MER's instruments were designed as a system that is coordinated to visually document and analyze the chemistry of rocks and soil at microscopic and panoramic scales. Nevertheless, the essential pattern of remotely operating a rover using visualization and daily science commanding was established with Sojourner, providing a proof of concept for MER.

Surface investigation requires new tools for navigation as well as a mission operations process that allows recoordinating the robotic system's operation on a daily basis, as Matijevic explains:

> When you are in the situation where you're trying to do these missions on a rover, you actually *require* information that is generally not paramount when you're flying a spectrometer or an imager or something around or by the planet. Here understanding the engineering that's possible with the device is essential for the scientists to acquire the information they are interested in. In other missions [the operations plan] becomes part of the initial design and once the instruments and the vehicle are deployed, either as an orbiter or flyby, you're actually *clocking* through the mission at that stage. The things that you do with the spacecraft are kind of well understood, even at the various design review stages, many years before the mission, before the instrument is ever deployed. In a rover mission, you are effectively creating "design behaviors." And now you're plopped down in an environment about which you have little if any information, and given the charter to go forward, do the exploration, find the opportunities to use your instruments. That requires a real understanding both of the engineers of the scientists and the scientists of the engineering: What are the capabilities and what is possible to be done?

The twin MER missions each initially cost about $400 million and were in planning through a laborious set of proposals over nearly a decade. The principal investigator, Steve Squyres, brought the team together through a complex obstacle course of NASA programmatic changes, in large part caused by the twin failures of Mars orbiter and lander missions in 1999. The team, originally called Athena, a name retained internally during the mission, often learned of each other's capabilities by competing for earlier canceled or revamped programs. They merged subteams with innovative instrument technologies and incrementally developed a conception they called a "robotic geologist" that could enable them to perceive and act on Mars, which they characterized as being a "surrogate."[15]

FIGURE 2.1/PLATE 3
Map of Spirit's traverse from the base of the Columbia Hills (near Hank's Hollow) to the summits, driving backward with five wheels and requiring more than one Earth year. *Image credit:* JPL, cropped from bottom. *Source:* JPL MER Mission, "Spirit: Detailed Traverse Map (August 15, 2005)," http://marsrovers.jpl.nasa.gov/mission/tm-spirit/spirit-sol572a.html (accessed January 31, 2008).

When Squyres said he wanted "to read the story of the rocks," he meant that he wanted an exploration system that would enable the scientists "to do real field geology on Mars"[16]—that is, the robotic tools would enable the team to do planetary field science by operating a mobile laboratory remotely from Earth. For example, describing the origin of the RAT, Squyres said, "Our rover was supposed to be a robot field geologist. When you see field geologists on Earth, they've got their boots, they've got their backpacks, and always, they've got big rock hammers."[17] Correspondingly, the micro-imager is like a powerful hand lens, the wheels can be programmed to dig trenches (like scraping your boot in the dirt), a brush on the RAT can sweep away dust, Hazard Avoidance Cameras (Hazcams) provide wide-angle ground views (120 degrees), and the Panoramic Camera (Pancam) and Navigation Cameras (Navcams) provide stereo images from a mast about 1.5 m off the ground (figure 2.2; shorter than the human height originally desired).[18] The RAT, brush, and Microscopic Imager (MI) are mounted on an arm with an elbow. Its reach is about the same as a human's (figure 2.3).

Despite these analogies between the rover and a geologist's tools, three additional, highly sophisticated instruments for detecting iron compounds and analyzing chemical element composition—Mössbauer Spectrometer (MB), Miniature Thermal Emission Spectrometer (Mini-TES), and APXS—make each MER more like a robotic laboratory. Geologists do not take such instruments into the field but instead analyze their rock and soil samples in laboratories at home (figure 3.3). Using such tools requires hours of "integration" for

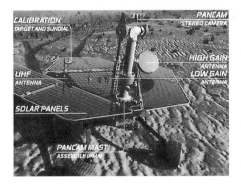 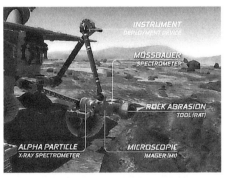

FIGURE 2.2
MER configuration and instruments.
Graphic: NASA/JPL.

FIGURE 2.3
MER Instrument Deployment Device (IDD)
and instruments. *Graphic:* NASA/JPL.

these sensors to "read" landscapes and materials on different scales. For example, Mini-TES pixelates a frame of view by individually programmed readings, moving a mirror to make multiple scans of an area, typically requiring five to thirty minutes but sometimes more than an hour. APXS integration could require ten hours. Such programmed analysis may require considerable power and time, which are both key constraints in daily planning.

The instruments allow the manipulation and analysis of materials in different ways that together enable useful interpretations (figure 2.3); accordingly, the design enables coordinated operations. For example, the Mini-TES has an internal telescope that shares the shaft of the Pancam mast and mirrors, so thermal scans and filtered stereo images can be correlated (figure 6.10).[19] Also, areas for close-up geochemical analysis are first scraped or brushed using the RAT to form a smooth surface, following from experience on Viking and Pathfinder, in which dust interfered.[20] The colocation of the Mössbauer, APXS, and MI on the Instrument Deployment Device (IDD) also facilitates placing them directly against common rock or soil targets.

Campaigns and Traverses

Having established the overall nature of the mission, it is useful to review how the exploration of Mars using MER has proceeded. Given these tools and objectives, how did the scientists use the roving ability to explore the Gusev and Meridiani regions?

The general pattern for operating the rovers has been to traverse to a large area of interest (a crater, summit, or plateau like Home Plate) and then remain in that area for some time doing a systematic investigation (tables 2.2 and 2.3). The traverse to the Columbia Hills required only three Earth months, but the investigation of Home Plate, an area of about 90 m in diameter behind the hills, persisted for several Earth years because of evidence of water in the presence of volcanism (and later limitations in Spirit's driving capability). During this time, the core team of about fifty scientists was studying an area somewhat larger than an American football field. These facts are useful for plotting a strategy for the future exploration of Mars (see chapter 10).[21]

The study of Endurance crater, about 130 m wide and more than 20 m deep,[22] lasted nearly six Earth months. The subsequent 7.5 km traverse to Victoria crater required twenty months, which is about 12.5 m/sol (41 ft/sol). The investigation of Victoria, prefaced by a traverse around its rim, persisted for over sixteen months, in a region about 750 m wide by 70 m deep. The rover was inside Victoria for over 340 sols, exiting August 28, 2008.

The next destination established for Opportunity was Endeavour crater (about 22 km in diameter, or 14 miles, 28 times larger than Victoria), about 19 km away. During the first 97 sols of this traverse (through December 2008), the rover drove about 1.8 km, about 19 m/sol (61 ft/sol), with the team stopping to study features found along the way. It was then driven 5.3 km (3.3 miles) in 2009, about 15 m/sol (49 ft/sol), again with periodic stops. In 2010, the travel distance was 7.6 km (4.7 miles), about 22 m/sol (72 ft/sol), indicating longer drives and less stopping. Indeed, the pace picked up even more in 2011 as the rover

TABLE 2.2

Major campaigns for Spirit at Gusev crater*

CAMPAIGN PHASE	START DATE (UTC**)	START SOL	DURATION (# SOLS)	DISTANCE (KM)
GUSEV LANDING SITE	JAN. 4, 2004	1	35	—
TRAVERSE TO BONNEVILLE CRATER	FEB. 8, 2004	36	30	0.35
BONNEVILLE CRATER	MAR. 10, 2004	66	21	—
TRAVERSE TO COLUMBIA HILLS	APR. 1, 2004	87	69	2.9
COLUMBIA HILLS (BASE)	JUNE 11, 2004	156	34	—
TRAVERSE TO HUSBAND HILL SUMMIT (HILLARY)	JULY 16, 2004	190	429	1.7
HUSBAND HILL (SUMMIT AREA)	SEPT. 29, 2005	619	33	—
TRAVERSE TO HOME PLATE	NOV. 2, 2005	652	94	0.85
HOME PLATE	FEB. 7, 2006	746	1464***	—

Notes

All data are extracted from JPL MER website "Rover Updates," http://marsrovers.jpl.nasa.gov/mission/status.html (accessed January 24, 2008). JPL updates specify the date of a status report, and do not generally correspond to the sols indicated. Correspondence between sols and earth time was calculated using the Mars24 program by Robert B. Schmunk. For example, Spirit landed about 2:35 p.m. local Mars time (Gusev) on sol 1, which was 4:35 UTC on January 4, 2004, or 8:35 p.m. PST (Pasadena, CA) on January 3, 2004. Traverse distances are approximations calculated from total odometry when available and incremental drives are indicated in the updates. Arrival and departure dates are only accurate within a few sols because the destinations are regions, not specific locations (except Hillary).

*"Start sol" represents approximate departure date to begin a traverse or arrival date at the designated area; durations at craters include investigations outside the crater before and/or after investigation inside the crater; Earth dates correspond to local Mars sunrise on sol indicated. Traverse distances are rounded to two significant digits.

**UTC = Coordinated Universal Time.

***The last transmission from Spirit was on March 22, 2010, sol 2210.

TABLE 2.3

Major campaigns for Opportunity at Meridiani Planum (key same as table 2.2)

CAMPAIGN PHASE	START DATE (UTC)	START SOL	DURATION (# SOLS)	DISTANCE (KM)
EAGLE CRATER	JAN. 25, 2004	1	69	—
TRAVERSE TO ENDURANCE CRATER	APR. 4, 2004	70	46	.75
ENDURANCE CRATER	MAY 22, 2004	116	200	—
HEAT SHIELD AND METEORITE	DEC. 13, 2004	316	42	—
TRAVERSE TO VICTORIA CRATER	JAN. 25, 2005	358	594	7.5
VICTORIA CRATER	SEPT. 28, 2006	952	702	—
TRAVERSE TO ENDEAVOUR CRATER	SEPT. 23, 2008	1659	1022	21.5
ENDEAVOUR CRATER	AUG. 9, 2011*	2681	?	?

Notes

Opportunity landed about 1:34 p.m. local Mars time (Meridiani) on sol 1, which was 5:05 UTC on January 25, 2004, or 9:05 p.m. PST (Pasadena, CA) on January 24, 2004.

*On sol 2681 (August 9, 2011), Opportunity's total odometry was 33.5 km (20.8 miles).

approached its target, with 5 km covered in the first six months (about 28 m/sol). Opportunity reached the rim of Endeavour crater on August 9, 2011 (sol 2681), after driving 21.5 km (13.3 miles; average 21 m/sol)—including driving backward since February 2009 (sol 1800) because of a right front wheel steering problem that developed in 2005.

By May 2009, it was evident that Spirit was mired in the sand. It was designated a "stationary research platform" in January 2010, after six years of operation.[23] The unfavorable position for power and hence insufficient heating of the electronics during the winter apparently led to its demise, with the last signal received March 22, 2010.[24]

As of the fourth anniversary of Opportunity's landing, the two rovers had taken 206,634 images together (table 2.4), plus three images each during entry and descent. Over the course of the mission to date, about 20,000 images were taken per rover for each Earth year. The average decreased from about 25,000/yr per rover during the first four years to about 8,000/yr for Spirit (averaged over the subsequent 2¼ years) and 14,000/yr for Opportunity (averaged over 4 years), reflecting the greater number of images required for Opportunity's navigation and Spirit's limited mobility offering fewer new features to study. The overall decrease in imaging rate is imposed by eliminating overnight downlinks; as the batteries have degraded in efficiency over time, it has become increasingly difficult to retain sufficient charge overnight for a middle-of-the-night transmission.[25]

By comparison, in 1997 Sojourner took 550 photographs and its APXS was applied to sixteen locations within 15 m of the landing site during eighty-three sols. MER's nominal engineering objectives for "full mission success" were to drive 600 m for the two rovers combined during a period of 90 sols, and for each rover to investigate four locations, take one color and one stereo 360-degree panorama, and use all of the instruments (including one RAT operation, image, and complementary observation). These criteria reflect the rover's capability to move and operate its instruments, rather than the scientific objectives.

TABLE 2.4

Image totals for the rovers on anniversary of Opportunity's landing

CAMERA	SPIRIT		OPPORTUNITY	
	JAN. 24, 2008	FINAL	JAN. 24, 2008	JAN. 24, 2012
NAVCAM	23512	27432	20653	36571
FRONT HAZCAM	6519	7432	5915	8846
REAR HAZCAM	2820	3351	2907	4494
PANCAM	68437	80568	65628	98958
MICROSCOPIC IMAGER	5387	6053	4856	7472
TOTAL IMAGES	106,675	124,836	99,959	156,341

Note

Data provided by Justin Maki, MER imaging scientist, in email messages to MER Science Operations team, January 25, 2008, and to author, January 30, 2012.

Timeline chart of MER mission events (2001–2011)

MER Mission Surface Operations

- Simulated Multiple-sol Operations using FIDO in "Mars Yard" (March-April) and Blind Desert Tests (May) 2001
- Preliminary Design Review – April 2001
- Critical Design Review - December 2001
- Operations Readiness Test (ORT) – August 2002
- Science Operations Training – Software: January; ORTs: August - November 2003
- Spirit Launched June 2003
- Opportunity Launched July 2003
- Prime ("Nominal") Mission – first 90 sols
- Solar Conjunction
- Southern Winter Solstice
- R9.2 Software Upgrade (dust devil & cloud detection; VTT; revised AUTONAV)
- Dynamic path planner and Visual Target Tracking tested
- Dust Storms Delay Operations – July/August 2007
- R9.3 Software upgrade
- AEGIS target selection used

Spirit Surface Operations

- Bonneville crater – March 2004
- Traverse to Columbia Hills – April to June 2004
- Columbia Hills
- Climb to Summit – July 2004 to September 2005
- Husband Hill Summit – October 2005
- Descend to Home Plate – November 2005 to February 2006
- Home Plate – February 2006 to March 2010
- Right front wheel stops working; driven backwards hereafter ->
- Sandtrapped at Troy – 6 May 2009
- Last comm 22 March 2010

Opportunity Surface Operations

- Eagle crater – January to March 2004
- El Capitan – February 2004
- Endurance crater – May to November 2004
- Right front wheel permanently stuck at 7 degree angle
- Stuck in Purgatory Dune – 26 April to 4 June 2005
- Erebus crater – October 2005 to March 2006
- Driven backwards hereafter ->
- IDD arm permanently deployed in front (unable to stow) ->
- Victoria crater – Sept 2006 to Sept 2008
- Endeavour crater

CHAPTER 2

The actual data returned demonstrates that the hardware, mission operations software, and team processes for controlling the rover provided capabilities for scientific investigation that far exceeded the minimal requirements.

Mission Timeline

To provide a further summary that will be helpful for understanding the extent of the MER mission challenges and accomplishments, figure 2.4 represents a timeline of events and activities. The graphic depicts preparation prior to landing, global events during the mission, and the primary areas of investigation for each rover.

Solar conjunctions, in which the sun is between Earth and Mars, prevent communication with the rover for about two weeks. During this period, the rovers run a preplanned sequence that can include applying instruments such as APXS and Mössbauer, but not moving the IDD or driving.

Around the time of the southern winter solstices, the rovers have less power because of the angle of the sun. Spirit is located at about −15 degrees south latitude, so it needed to be parked at an angle tilted to the north; Opportunity's location closer to the equator (at about −2 degrees) has enabled it to drive during the winter, with reduced activities during the week of the solstice when power is at the theoretical minimum. Dust storms such as the one in 2007 can have an even greater effect on operations. The variation in power during the course of the mission is complex, and engineers have adapted. For example, the 2007 dust storm cut Opportunity's power production by more than half to below 300 watt-hours/sol, which stalled operations; but during 2011, the rover continued its trek to Endeavour despite accumulated dust making 400 to 500 watt-hours the norm.

The importance of Home Plate and Victoria for the scientific investigations is evident in the chart from the time devoted to those areas. The parallel activities at those locations underscore how well the complexity of two missions was handled and also the intensity of the activity during the martian summer of 2005. The long effort of traveling to Endeavour in 2009–2011 is also striking.

With the mission background and sketch of the inquiries in mind, we are prepared to understand in what respects the science operations are new and what we have learned about planetary field science. What is field science? How does it relate to exploration? What are different ways of using robotic tools, and what determined and enabled the daily commanding method used on MER? The next two chapters examine these questions.

FIGURE 2.4
MER mission timeline, including surface operations preparation, global events, and individual rover investigations.

3

A New Kind of Field Science

On an overcast July afternoon on Devon Island in the Canadian North Arctic, I was climbing a scree slope of loose, ankle-bruising rocks with a geology graduate student. Ahead on a sheer cliff wall was an outcrop of interest (figure 3.1/plate 4). Its horizontal bands, broken by a fault, revealed the sedimentary undersea past of this valley. As we were sliding and struggling for footing, I remembered how Apollo 15 astronauts avoided descending into slopes at Hadley Rille and wondered whether Mars astronauts would be allowed to take such risks. Or could we design a rover that could climb like this?

In my role as "participant observer," I took photos for the geologist. I wondered how the robot would position its camera close against the rock's edge, 5 feet above the unstable slope. And if this were a camera used by an astronaut, could one draw directly on its display to annotate the image? And when would you do that—standing here, sitting on your Mars all-terrain vehicle, or back in the habitat?

Besides learning about geology and imagining implications for a Mars mission and robots, I was learning about geologists, too. My initial request to come along revealed that this geologist was looking forward to having this time alone. But the expedition's safety rules required traveling at least in pairs. Maybe he would have preferred a robotic assistant?

This afternoon's traverse was just one of dozens by geologists and biologists who crisscrossed the 20-kilometer-wide Haughton Crater accompanied by computer scientists and robotics engineers during that summer's "Mars analog" expedition. It was such experiences that made the science assessment meeting at JPL in its darkened room (figures 1.1 and 5.5) appear so strange. Simply put, the MER mission is the first time people have conducted field science on another planet with a programmable robotic laboratory. To understand what this means—what the MER scientists are doing on Mars, how such remote work is possible, and how this mission is different from others—it is useful to learn a bit about the practice of field science on Earth.

Exploring a Place Scientifically

Field science is an investigation of some place (the "field"), focusing on the land, climate, artifacts, and other aspects of that area, according to disciplinary interests. Field science most often requires travel and temporarily living in or near the area being studied, sometimes as individuals but more often on group trips called *expeditions*. Such trips are often experienced by the scientists as being adventurous experiences in which they explore a place for the first time, document it, and seek to grasp spatial and temporal relationships of new kinds of things and phenomena. Their excitement in exploring has been captured historically in travel journals (and today may be recorded and shared during the journey in online blogs). Participants may feel an elevated sense of being alive and of their personal importance: early "glaciologists and other field scientists identified themselves with heroic explorers."[1]

In some disciplines, such as geology, field science plays a primary role in defining what the scientists study.[2] Field scientists identify what phenomena need to be understood by seeing what is out there and gathering comparative data across times and places. Such investigations may occur almost anywhere on Earth, paradigmatically illustrated by these disciplines (with examples of what they study in the field):

- Biology/ethology (e.g., jungle species, extremophiles in caves)
- Geology (e.g., lava fields, southwest US desert formations)
- Meteorology (e.g., ice cores, tree rings, Antarctic weather)
- Anthropology/ethnography (e.g., third-world cultures, office workers)
- Paleontology/archaeology (e.g., African rift zone fossils, Mesoamerican villages)

In these disciplines, field science is a skill usually taught in university courses by making field trips (also referred to as "doing fieldwork," frequently offered to undergraduates as a "summer field camp"). The learning process may involve a paid apprenticeship assisting a professor's funded research projects, just as student interns participate in planetary science missions. Doctoral dissertations typically involve a student defining and carrying out fieldwork in a new place and/or from a new analytic perspective.

FIGURE 3.1/PLATE 4
Geologist in Haughton Crater on Devon Island on a scree slope inspecting fractured surfaces during the Haughton–Mars analog expedition, July 1999. Field geology is often done alone or in small groups, climbing to reach outcrops and survey surroundings. *Image credit:* NASA Haughton–Mars Project/Clancey.

The creation of field science as an activity involved a shift in the nature of science itself: getting to know a place versus placeless truths. Kohler explains, "Natural places are particular and variable places, none quite like another, each the result of a unique local history ... unpredictable, unrepeatable, beyond human control."[3] Places are unlike objects of study such as magnetism or light, processes whose properties can be described by universal laws. The first goal of a geologist in studying a place is to know its character, to determine its particular history.[4] For example, what are the rock layers in Eagle crater, and secondarily, are they related to those in Victoria crater? Each place must be first studied locally, cognizant of global processes and general geological, climatological, and chemical processes. Afterward, the individual biographies of the craters, hills, deposits, and other landforms can be tied into a broader story about Mars.

The "particularity of place" is certainly important for understanding ecologies and cultures but somewhat less valid for landscapes, in which the general force of water, atmosphere, and tectonics deposits, erodes, and shifts landscapes in recognizable ways. Indeed, everything known about geology and meteorology on Earth provides a baseline for understanding what we see on the moons and other planets of our solar system. (We apply biological experience similarly, though with some hesitation.) In particular, Earth analogs are frequently important in identifying martian formations: stories about what happened on Earth can be useful for creating stories about Mars's past. A prominent example is how the formations on Mars have been related to features on Earth caused by water, wind, volcanic, and impact processes.[5] Two cognitive steps are involved in drawing such analogies: perceptually recognizing similar morphology (e.g., a crater shape, layers, valleys) and identifying the causal process that created or modified the structures over time (e.g., an impact, wind deposits, a catastrophic flood). Of course, a great deal of empirical work—perhaps involving detailed geochemical and atomic analyses—may be involved to convert a superficial analogy into a well-supported theory, such as the work involved in determining different causes of layers found in Eagle and Victoria craters.

Analog Research and Operations Tests

Given the role of analogies in understanding planetary surfaces, continuing field science on Earth plays a central role in advancing our understanding of Mars. Guided by the needs and funding of the space program, field science on Earth has become a multidisciplinary activity that prepares for planetary field science by studying "analog sites." For example, the region of Haughton Crater on Devon Island provides an analog for martian periglacial terrain (figure 3.1). Field science at analog sites often involves experimenting with operations protocols and prototype technologies that may be useful on the moon or Mars.[6]

A variety of expeditions to analog sites, called "analog missions," have been funded by NASA and the Canadian Space Agency since the late 1990s. In these projects, scientific work has determined the expedition's location, and operations and technology research revolves around and supports the scientists' investigations. Examples of analog missions focusing on geology and/or extremophiles are (indicating location, scientific analog of interest, and exploration operations/technology analog aspect):

- Haughton–Mars Project (Haughton impact crater on Devon Island, Canadian Arctic; Mars periglacial formations; telemedicine, robotics, and mobile habitat)[7]
- Marte Project (Rio Tinto, acidic/geothermal river in Spain; anaerobic extremophiles; drilling with remote science team)[8]
- Pavilion Lake Research Project (Canadian lake with carbonate rock structures believed to be formed by microorganisms; ancient fossils on Earth; underwater vehicles)[9]
- Licancabur Expedition (highest freshwater lake, in Chilean Andes; evolution of paleohabitats; high-altitude diving)[10]

Other analog missions have emphasized operations experiments with technology prototypes, usually in difficult operational environments, such as the Atacama Desert Trek[11] (mobile laboratory in Chile with remote science team); the ENDURANCE Project (underwater robotic investigation in Lake Bonney, Antarctica, an ice-covered sea serving as an analog for Europa);[12] Desert-RATS (testing pressurized suits, rovers, and robotics near Meteor Crater and lunar volcanic sites in Arizona); the Mars Desert Research Station[13] (living and working in Mars analog habitat in Utah); and NEEMO (undersea habitat and diving operations in the Florida Keys).

Michael Carr, one of the MER scientists, explained why studying Mars analog sites on Earth is especially relevant for planetary scientists: "The thing about Mars is that you can take your terrestrial experience and transpose it there. It's sufficiently different to make it interesting. It's not the Earth. But all the same processes are going on, so you can have some sort of gut feeling because of your terrestrial experience. When you go to places like Io and Titan, you don't have that. It's different. You don't feel at home there. Whereas on Mars, it's a lot like the Earth."[14]

Analog mission operations have included rehearsals of remotely controlled robots with distant science teams, constituting imaginative rehearsals for future missions. In developing MER's surface operations, and particularly the field science process itself, simulated operations were undertaken in operational readiness tests (ORTs) using a functionally similar MER prototype called Field Integrated Design and Operations (FIDO).[15] These integrated tests of the planned operations for MER experimentally exercised the communication protocols and tools that related the science and engineering activities. They tested the ability of scientists to interpret and guide the work through MER's instruments, just as the engineers were tested in their ability to control the rover from photographs and telemetry and their programming tools (see "Learning to Work Together during Analog Field Experiments" in chapter 7). Analogous "science end-to-end tests" were conducted for Viking to assess the ability to carry out scientific investigations "from sample collection to interpretation of resulting data by the scientists" and to familiarize people with the programmable systems.[16] However, for MER the simulation also involved actual field investigation: FIDO was placed in a realistic field site unknown to the science team, and the team was challenged to conduct fieldwork entirely through the sensors and effectors of the robotic laboratory.[17]

Fieldwork Methods

To appreciate the design of MER's technology and operations, as well as its limitations, it is useful to understand the nature of field science on Earth in a bit more detail. First, fieldwork usually involves observing by being physically present and creating location maps (e.g., land surveys) of the physical layout of the phenomenon of interest (e.g., soil, plants, buildings, tools, people). Scientists choose a general area to investigate and then destinations and routes within it. They select tools, plan schedules, and assign roles and tasks to participants. Fieldwork involves moving and looking. Geologists refer to a particular outing ranging over several hours to a day as a "traverse" (e.g., walking or driving to a land feature such as a hill with outcrops). The descriptive aspect of fieldwork is characteristic of biology, too. For example, comparing laboratory and field biology, Kohler says, "Experimenters analyze and reveal causes and effects; field biologists more often describe, compare, name, classify, map. In fieldwork spatial and locational ways of knowing have equal standing with causal reasoning."[18]

Field scientists make notes about observations and interpretations, usually in a special field notebook (figure 3.2). They emphasize the importance of studying the context in which an interesting phenomenon occurs (e.g., where a rock is found) and the relative layout of features (e.g., do the layered deposits we see here occur somewhere else along the horizon?). Such surveys usually involve taking photographs and samples (figure 3.3/plate 5) and logging their locations on maps. More recently, electronic recording instruments provide opportunities to better document previously studied sites (e.g., using LIDAR [Light Detection And Ranging] to profile lava tubes). Field scientists use maps extensively and may refine existing maps, but cartography as a discipline is not itself field science.

Field science is inherently exploratory in seeking to know and explain a setting: the work is opportunistic, serendipitous, and incremental. The work of an expedition is highly contingent on what the scientists find as they go along: what is worth investigating in detail? What patterns are emerging that should be mapped and related? During their travels, the scientists may find and indeed hope to stumble upon something unexpected (e.g., the orange soil that so excited the Apollo 17 astronauts). Landing in Eagle crater is a fine example of such serendipity (figure 3.4). Cognitively, fieldwork exemplifies "situated action," in which plans are adapted, refined, and reworked and details are improvised according to broad objectives as more information is gained and new interpretations come to mind. In conventional fieldwork, most of the detailed analysis occurs back at the

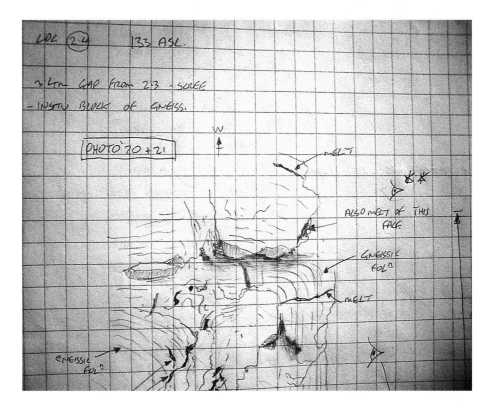

FIGURE 3.2
A geologist's field notebook often includes
drawings made on location, with measurements,
photograph numbers, and interpretations,
here an "in situ block of gneiss." *Image credit:*
Gordon Osinski/NASA Haughton–Mars Project/
Clancey.

FIGURE 3.3/PLATE 5
Samples collected by a single geologist in Haughton Crater during
one week (July 1–8, 1999). Bag labels are referenced in written notes.
A selection from the monthlong expedition was shipped back to
the geologist's university laboratory, where thin sections were polished
and investigated using a digital scanning electron microscope; none
were analyzed in the field. See Gordon Osinski, John Spray, and Pascal Lee,
"Impact-Induced Hydrothermal Activity within the Haughton Impact
Structure, Arctic Canada: Generation of a Transient, Warm, Wet Oasis,"
Meteoritics & Planetary Science 36 (2001): 731–745. *Image credit:* NASA
Haughton–Mars Project/Clancey.

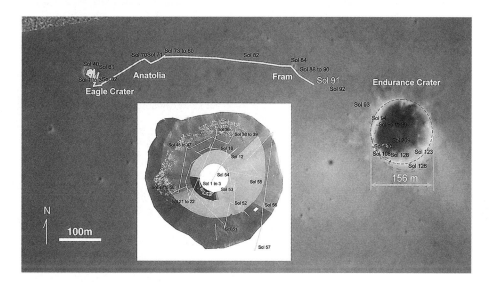

FIGURE 3.4
Traverse of Opportunity within Eagle crater and then to Endurance. The scientists' choices of route and durations illustrate the nature of field science as intensive investigation of particular places (fifty-six sols in Eagle crater) followed by veritable sprints to highly anticipated features with slight jogs to stop briefly at places of interest along the way. *Image credit:* Squyres et al., "The Opportunity Rover's Athena Science Investigation at Meridiani Planum, Mars." *Source:* Data supplement plate 5: http://www. sciencemag.org/content/vol306/issue5702/images/large/1715-5-large.gif (accessed July 5, 2009).

scientists' home institution, particularly analysis of samples using laboratory instruments—involving subsequent data correlating and charting—and writing detailed reports.

The iterative process of scientific fieldwork typically extends over multiple field seasons in different years. Carol Stoker, a Pathfinder participating scientist, has used her field experience to develop tools for planetary fieldwork and to formulate concepts for remote scientific exploration. Her anticipation of how new topics of study will arise during field science on Mars is instructive:

> Key areas of scientific inquiry and science priorities are identified at a high level. Scientists propose individual investigations which usually result in field teams visiting a particular site for several months of each year, for three to ten years. By analogy, scientific fieldwork on Mars will require investigators returning repeatedly to field sites of special interest over a period of years. Furthermore, science interests are likely to evolve into new areas of study resulting from knowledge gained during the fieldwork.[19]

Field science, with its emphasis on grounding its questions and data on naturally occurring phenomena that are explored in place, can be contrasted with investigations based on laboratory experiments (e.g., high-energy physics; psychological testing; animal experiments) and surveys (sociology questionnaires). Historically, the relation of the field to the laboratory has been a source of conflict in biology, ethology, psychology, and the social sciences because of the inherent tension between finding complex, authentic phenomena and the scientific drive for defined (bounded and controlled) events, precise measurement, and well-articulated causal models. Kohler explains that this inherent tension has led opposing methodologies to "co-invent" each other; that is, they are codependent—they define and adapt to each other.[20]

For geologists, there is less choice. Biologists can grow plants and animals in the lab. But in geology, landscapes are the subject and must be characterized in place. Investigation of strata, uplifts, buttes, sedimentary deposits, and other landforms occurs first in the field. Laboratory work may simulate geological processes (e.g., how differential sorting of materials occurs through wind or water). But lab work also becomes a subdiscipline, such as geochemistry. Thus volcanologists studying a lava field, following and documenting the forms and dynamics of flows, may not know or care about the particular chemical compositions. This distinction between field and lab science interests is important in understanding the collaborative work on the MER team (see "Field versus Lab Scientists: Disciplinary Conflict from Pace" in chapter 7).

The laboratory-field tension becomes salient on MER because field science ultimately requires both settings, and given the inability to return samples, instrument-based laboratory analysis must be done on Mars; this causes a traverse to be stopped while the quality of data is determined. In this manner, initial analyses help guide the ongoing choice of targets in the local area, and the search for related or complementary features and strategically orient the ongoing route for the investigation. Indeed, although we now take it for granted when we talk about MER, it is remarkable and strange from an Earth fieldwork perspective that the entire martian laboratory must be trundled along on each traverse—up and down hills and crater walls and then around rocks and slippery sands. The mixture of observing and analyzing by different field and laboratory disciplines in a collaborative field science investigation is therefore somewhat bizarre from the perspective of Earth practices. Some team members focusing on instrument analysis might never have participated in large-group field expeditions. And some geologists never used the kinds of instruments—either in the lab or in the field—that they are now using to guide their work on the martian surface.

Relation of Laboratories and Planetary Science

Historically, until laboratory work became institutionalized in the twentieth century, not all instruments "belonged" in a laboratory. Indeed, Kohler tells us, "Before the era of laboratory building, instruments and measurements were not necessarily thought of as peculiar to laboratories or as items to be 'borrowed'—they were of the field."[21] But with MER, scientists are taking unusually expensive, institutionally owned laboratory equipment—refined or customized for operating on the cold surface of Mars—to the field. Instruments like the MI and the APXS would normally be applied only to samples brought back to the laboratory, not analyzed in situ, partly because of the instruments' size and weight and also because they are usually not individually owned or free to be borrowed. On the other hand, cameras are usually field-ready, shareable tools but unlike MER's cameras are not often programmable. The RAT lies in between, somewhere between a hammer or brush for abrading the surface (which a geologist would carry) and a tabletop polishing lathe (as might be in the work tent at Haughton Crater); similarly, the MI lies between a hand lens and a microscope.

Because MER samples are analyzed in situ and are not brought back to Earth, the measurement precision must be as good as possible—that is, meeting laboratory standards, including being able to specify and document where measurements occur. Just as use of instruments on unpaved rural roads early in the twentieth century required hardier design, MER's instruments have been designed for the launch, the long cruise, and the jarring and cold cross-country travel on Mars.[22]

Of course, some of the characteristics of MER's instruments are common to planetary missions in general. Orbital and flyby missions (e.g., Voyager, Cassini, Mars Reconnaissance Orbiter) and landers (e.g., Viking, Huygens) all bring scientific instruments to the place that is being studied. All involve teams of scientists who do not use such equipment in their own laboratories or in the field. All have some relation to field science by virtue of enabling people to study a remote place. The essential difference is that orbital, flyby, and lander missions do not enable studying the surface by navigating through the landscape (avoiding and struggling to traverse different formations) and physically manipulating its

FIGURE 3.5
Traverse map of Spirit from arrival at Columbia Hills (upper left), climb to
the summit (area labeled sol 600), and eventual investigation of Home
Plate (loop at bottom). *Image credit:* NASA/JPL/Cornell/MRO–HRISE/NM
Museum of Natural History and Science. *Source:* JPL MER Mission,
"Spirit: Detailed Traverse Map (4 Jan 2008)," http://marsrovers.jpl.nasa.gov/
mission/tm-spirit/spirit-sol1424.html (accessed January 21, 2008).

features (e.g., opening rocks, cleaning outcrop surfaces, drilling). Viewed from afar, the missions seem superficially identical: the Cassini mission enables the study of Titan (and the other moons of Saturn); MER enables the study of Mars. But the Cassini mission does not allow scientists to study Enceladus as they might on the surface; Huygens, designed primarily to study Titan's atmosphere, gave only the barest glimpse of the landscape.

In some respects, the difference between an orbiter (like Mars Global Surveyor) and a rover (MER) is akin to the dramatic difference we have all experienced between flying over a landscape and walking through it. Mike Carr, an astrogeologist who has now had extensive experience with both modes of studying Mars, finds the shift jarring, even difficult to negotiate—working from orbit and working on the surface using a rover require different ways of thinking.

Landers such as Surveyor, Viking, Huygens, Venera, and Phoenix (table 2.1) of course enable studying a planetary surface. But standing in one place, whether for an hour or several years, does not constitute field science. What is missing is getting to know a place by exploring the landscape, digging and manipulating a variety of places, and seeing what's over the next hill so that you can understand particular features by studying the surrounding context and relate neighboring regions to one another. And this is what the scientists are doing on Mars—repeatedly surveying possible areas to explore, looking closely, and then determining new places to go, ranging from craters to outcrops to a palm-size area of rock. This kind of inquiry is characteristic of field science.

Humboldtian Science: Systematic, Quantitative, and Integrative Inquiry

How people think about MER, and in a large part how we think about space exploration, weaves specific scientific details and interests with historical allusions and motivations. In one formulation, which I will discuss subsequently, historical references support broad—often nationalistic—aspirations for a space program. In another formulation, historical references relate to the roots of field science and a way of thinking that illuminates MER's design and the scientists' strategy for understanding Mars. In particular, we consider here how early field science in the eighteenth and early nineteenth centuries by European explorers relates to how MER scientists frame their investigation, both historically and methodologically.

The MER scientists' reveal their historical perspective by how they are naming features on Mars. Symbolizing an affinity and identification with European voyages of discovery, Jim Rice advocated naming craters after famous ships and places visited during historic expeditions. The team has informally named Victoria crater's "bays" and "capes" after landmarks visited by Ferdinand Magellan on the ship *Victoria*.[23] Fran, Endurance, and Resolution are other examples at Meridiani. In naming Endeavour crater's features, the team uses names of places visited by James Cook in the 1769–1771 Pacific voyage of the *HMS Endeavour*.[24] In this way, the scientists weave into their rigorous scientific work a playful, somewhat romantic view of the MER expedition. (Formal names are assigned by the International Astronomical Union; "Martian craters are named after famous scientists and science fiction authors, or if less than 60 km in diameter, after towns on Earth."[25])

Besides a romantic idea, the analogy between MER and early voyages of discovery has some depth. The relation is of course often repeated in describing the space program. For example, NASA's classroom *Liftoff to Learning* series,[26] "Voyage of Endeavour: Then and Now," compares the scientific work onboard the Space Shuttle Endeavour to Cook's geographic exploration.[27] The analogy is more salient for MER's field science, which—like early exploration on Earth—involves traveling through, surveying, and documenting largely unknown, remote, and dangerous lands on a multiple-year journey. Of course, the existence of a place may be known, but little might be known about it. The Pacific Ocean was known to the Europeans for more than 250 years before Cook's journeys, but its extent and the lands bordering it were largely a mystery.[28] The investigation of the Columbia Hills (figure 3.5) and Home Plate (figure 10.1), although not on a continental scale, involved first-time surveys of the landscape. Similarly, Victoria crater was visible from satellite images, but before our overland arrival, the basic morphology of the rim was unknown. On a small scale, the experience of the traverse was like exploring a coastline—Capes St. Vincent and St. Mary (figure 5.4) were named after prominent sailors' landmarks in Portugal.

In effect, MER illustrates field science with an exploratory aspect (e.g., detouring Opportunity to investigate a Santa Maria crater on the way to Endeavour crater); for Cook's voyages, the emphasis was reversed—field science occurred as a "carry-along" activity during a predominately early mapping and navigation effort. In this respect, the failure of Cook's second voyage to find Antarctica and publicized erroneous conclusion that it did not exist presents a fair warning in interpreting our early surface investigations of Mars. Confined to Spirit and Opportunity, we have little idea what surprises are hidden—perhaps just meters away from their perambulations. Consider for example, the serendipitous discovery of the white scuff marks that turned out to be almost pure silicon near Home Plate. What lies over the hills around Spirit, perhaps providing the very answers we seek about the early history of Mars and origins of life on Earth? (And was there ice waiting to be discovered only 10 cm below Viking 2's shovel?[29])

The analogy between MER and early Earth expeditions can be significantly extended by considering the work of Alexander von Humboldt (1769–1859), a German naturalist and explorer. Humboldt articulated and demonstrated how exploration could be done scientifically: "Where earlier explorers had merely observed, the aim of the scientific traveler was to record more accurately, to measure more precisely, and to collect more comprehensively."[30] In effect, Humboldt laid the foundations for a field science methodology using instruments that has now been realized telerobotically on Mars. Humboldt would probably appreciate the MER scientists' coherent, multidisciplinary strategy, though—like the MER scientists themselves—he might be disoriented by the communal, public approach.

Humboldt's scientific exploration approach was inspired by Cook's expeditions, which he credited for including groundbreaking science by displaying "the potential of scientific expeditions."[31] In particular, Humboldt described the contributions of Georg Forster as beginning "a new era of scientific travels."[32] Georg (1754–1794) joined his father, German naturalist Johann Reinhold Forster (1729–1798), on Cook's second voyage, on which they "undertook extensive studies in 'physical geography, natural history and ethnic philosophy [anthropology].'... (The elder) Forster's work provided one of the most significant models for the Humboldtian style of exploratory science."[33]

In effect, Humboldt strived to carry out an integrative investigation through his own multidisciplinary interests, as is realized on MER through the collaboration of dozens of specialists from the science theme groups: "While undoubtedly following the eighteenth-century model of scientific exploration, Humboldt displayed a broader variety of scientific interests and intellectual concerns than any other explorer before or, indeed, since."[34] The MER team's integrative study of Mars fits the holistic, systems-perspective that developed throughout the sciences in the twentieth century. But such an idea was new in Humboldt's time. Following the philosopher Kant, Humboldt strived for an integrative study of "unifying processes that were not directly visible," assuming that the character of a given region developed from the interactions of all its phenomena.[35] This is also why MER was not designed as a consortium of payload principal investigators (PIs), but as a tool for an integrated team to study *the planet*.

Crucially, Humboldt's empirical method was realized by purchasing and learning to use a complement of about twenty different types of instruments for measuring everything from the blueness of the sky to the quantity of oxygen in the atmosphere.[36] Exploiting the latest advances in instrumentation (e.g., a magnetometer), Humboldt was able to transport and employ an effective portable laboratory including chemical reagents, though not without considerable cost and as many as twenty mules.[37] Thus, although Humboldt embraced the Forsters' adventurous expedition style, his instrument observations go beyond cartography and taxonomic description—for example, enabling studies of plant life and the climate to be systematically related.

In further parallel with MER, Humboldt's deliberately land-based exploration contrasted with coastal voyages, availing him with a greater extent of phenomena studied in a more sustained manner, "a pioneering demonstration of the feasibility of sustaining long scientific expeditions on land."[38] This continuous, sustained presence in a region echoes the MER scientists' laborious step-by-step roving for about two years from Endeavour crater to Victoria crater (and then two years in and around it), followed by three years on to Endeavour, and similar painstaking investigation with Spirit around Home Plate. Yet for Humboldt, as well as the MER geologists, spending years in an area of a few kilometers would feel tediously slow. (Archaeologists of course have a different expectation, commonly devoting years to an area measured in meters.)

Summarizing Humboldt's fundamental contribution to field science, Darwin wrote, "You might truly call him the parent of a grand progeny of scientific travelers, who, taken together, have done much for science."[39] With its systematic, multidisciplinary, and instrument-based observation, MER's voyage of scientific discovery fits into this tradition.

Understanding Science, Discovery, and Exploration

We are closer now to understanding two different views of MER's mission operations: the difference between what the scientists are doing on Mars and what the rover is doing in carrying out the scientists' observation plans. In particular, having established how the MER scientists are doing fieldwork on Mars and a bit about their way of thinking, we are now positioned to ground the discussion about "science" and "exploration" in a new way so that we can appreciate how robotic tools enable scientists to explore.

Despite hundreds of years in which scientists engaged in exploratory expeditions, the meaning of "exploration" has become confused in the space program community. Indeed, one historian of space exploration suggested to me that the MER was misnamed: because the area had already been surveyed from orbit, the landscape was already known, and hence no exploration of the Columbia Hills or Meridiani has been involved at all. Of course, this is not true; the scientists were quite surprised on arrival to find that Gusev was not covered by lake sediments as orbital data had suggested.[40] And a glance at Spirit's path before the Columbia Hills shows how the scientists detoured to Bonneville to explore what was there.

Sometimes even advocates aren't quite sure what words to use: the Planetary Society has argued for the integration of science and exploration, a combination of "discovery and adventure," as if someone could successfully separate them.[41] The society's mission statement appropriately states that science and technology are inseparable, but then presents the science fiction of "robotic and human explorers working together in an integrated program."[42] The words "science," "discovery," and "exploration" are tossed about in many statements like this, in which people are often advancing different goals and interests, especially in allocating funds between human spaceflight ("human explorers") and planetary science ("robotic explorers"), or as in the Planetary Society's case, supporting both.

The form of these puzzles is classic in philosophy: after treating what is logically a *both-and* dependent relation as if it is *either-or*, academics and lobbyists struggle to argue how or when the parts A and B—which they now presuppose could exist independently—should be brought together.[43]

Traditionally, "space exploration" has been appropriately used as an encapsulating phrase associated with all of NASA's missions, both planetary probes and human spaceflight. For example, an early advisory committee report to President-elect Kennedy in 1961 listed among the primary activities in space the "exploration of the solar system with instruments carried in deep space probes" and concluded that until people in space were able to accomplish scientific and technical tasks, "space exploration must rely on unmanned vehicles."[44] This use of the term "exploration" is particularly common in subsequent mission accounts, such as *On Mars: Exploration of the Red Planet: 1958–1978*.[45] Accordingly, the words "exploration" and "exploring" provide a convenient shorthand that most people understand as referring to the objective of the space program in general; as the acting NASA administrator told the US Congress in 2009, "NASA's astronauts and robotic spacecraft have been exploring our solar system and the universe for more than fifty years."[46]

In the context of geographic studies of a planetary surface, exploration is identified with *reconnaissance*, seeing new places and visual features. As explicated previously, field science is inherently exploratory in this sense, in an effort to make scientific discoveries about a place. Apparently with this meaning in mind, NASA historian Steve Dick said, "Exploration done properly is a form of science."[47] Humboldt made this quality explicit by using the terms together: "scientific exploration." The phrase also appears in NASA's mission histories (e.g., a chapter titled "Viking's Scientific Explorations"[48]).

Among scientists, in referring to the work of learning about a phenomenon, "exploration" is viewed quite broadly, including anything involving scientific observers. So, for example, astronomers are exploring the universe using telescopes, including spacecraft like the Hubble Space Telescope. Accordingly, physicist Freeman Dyson equates "human exploration" with "the search for the basic constituents and underlying laws" of the universe, ranging from subatomic physics to cosmology.[49]

Certainly, MER scientists are not confused about how to use the term "exploration." They experience the unfolding events as characteristic of exploring, meaning to look and learn about something—for example, upon examining a Navcam image on sol 1148, Squyres said about the amazing image of cross bedding, "It was completely fortuitous .... Exploration is like that."[50] Their publications similarly relate roving and investigating: "Exploration of Victoria Crater by the Mars Rover Opportunity."[51]

Confusion over the term "exploration" was formalized by "Vision for Space Exploration (VSE)"[52] in 2004, in which the term "exploration systems" was interpreted as referring to technologies enabling future human spaceflight (with the relation of people, science, and technology confused further by the reference to "robotic exploration" in the strategic objectives). The meaning of the words became institutionalized into a dichotomy in the following years, with a NASA "Science Directorate" focusing on planetary probes and an "Exploration Systems Directorate" focusing on human spaceflight.

With this institutionalized division, something confusing occurred: science and exploration were no longer conceptually or practically linked—as they occur naturally in human activities—but were now opposing camps, competing for a budget. The distinction between human spaceflight and planetary science then became an issue of public policy about "the balance between human and robotic exploration."[53] Rather than appreciating that people were already working through robots in planetary science, supposedly new questions were raised about how to "integrate" people with them.

Framed this way—referring to "the entity working out there," as if a programmed system existed independently—we have either a competition or an imaginary collaboration between "human explorers" and "robotic explorers." And now "to explore" means to *be physically there*. Thus, we have a distinction between "exploration by the physical presence of people" and "exploration by the physical presence of robots." In this vein, Stephen J. Pyne defines "the explorer" to be the remote entity, physically present in the region being investigated. He argues that people have a special role and experience in encountering other beings, as occurred in the exploration of the Earth. But "with no distinctively *human* encounter possible, there is no compelling reason for humans to even serve as explorers."[54] And now you see the pitfall: according to this analysis, the MER scientists conducting fieldwork on Mars are not exploring.

Referring to robotically programmed machines (remotely operated by scientists on Earth) as being the "explorers" has become engrained in all kinds of publications. For example, Reidy, Kroll, and Conway refer to "findings returned" by the MER and Cassini "robotic explorers."[55] *Science* journalists write, "The rovers enjoy considerable political support in Congress, and the public finds the intrepid explorers adorable."[56] The scientific work and the scientists are lost in this poetry. At least Chaikin's recent book title inadvertently plays with the ambiguity: *A Passion for Mars: Intrepid Explorers of the Red Planet*.[57] He says, "I always think of the robots as explorers."[58] But the title and most of the book is actually about the scientists and engineers of the space exploration community.

As illustrated in the first chapter in this book, even the MER scientists equivocate in attributing their work to the rover. At heart is a difficulty regarding knowing how to talk about the relationship between what the people are doing and what the rovers are doing—or indeed, the work they are accomplishing together. A 2003 JPL blog about the FIDO field tests entitled "The Human-Rover Partnership" claims, "Since humans cannot go to Mars yet, the Mars Exploration Rovers will act as robotic scientists." Presumably this means that the robots will perform as scientists, rather than pretend to do so. The next sentence idly belies the first: "But who are the *human* scientists calling the shots—both during the field tests and the actual mission?"[59] So much for the partnership.

In large part, understanding the nature of robots is complicated because we don't understand people well, either, as illustrated by the prevarication of another JPL online story: "People Are Robots, Too. Almost."[60] We appear unwittingly good at viewing people and technology interchangeably—the brain was imagined to be like a computer, and next people were imagining supercomputer–humans.[61] A respected Mars astrobiologist (not a member of the MER science team) reported in a conference plenary presentation, "The MER robots are really scientists; they are functioning as scientists."[62] Yet the machines have no scientific goals (let alone personal interests, including survival), no ability to use time wisely on any given day (let alone plan the mission strategically), no ability to decide what data to record (let alone perceive unexpected distinctions in the photos and measurements), and no ability to relate data to geological or biological hypotheses (let alone conceive new interpretations).[63] They function as mobile laboratories, not scientists.

What lies behind the ready social adoption of superficial metaphors? George Orwell's famous 1946 essay "Politics and the English Language" argues that the "slovenliness of the language" ultimately has economic and political causes, and accordingly, "Our thoughts are foolish." His first rule for ensuring that "the meaning chooses the word" is to avoid clichés: "Never use a metaphor, simile, or other figure of speech which you are used to seeing in print."[64] If the only word that ever follows "intrepid" is "explorer" in print, then the words are choosing the meaning. Building on Orwell's analysis, Edward Tufte's critique of the "slovenliness" of today's technical writing and presentations rings eerily familiar: "Like puns, these words have a duplicity of meaning, with narrow technical and broad cheerleading meanings." Not coincidentally, Tufte has singled out NASA's engineers for lacking training in clear speaking and perspicuous drawing, notably in his dissection of the Microsoft PowerPoint reports presented during Columbia's ill-fated mission.[65]

Sometimes the politically imagined antagonism between science and exploration is explicit, as in the title of the February 2008 meeting "Examining the Vision: Balancing Science and Exploration."[66] Or it may be implicit, as in the September 2004 "Risk and Exploration" workshop,[67] which in relating the risks of human spaceflight to mountain climbing, caving, and so on seemed to be almost an inquiry into why Sean O'Keefe, then NASA's administrator, canceled the Hubble repair mission (which eventually moved forward under Griffin's guidance in May 2009). Perhaps you can begin to see how using "exploration" as the middle name of a robot would strike some people as advancing an agenda and might even be confusing.

Distinguishing Sociohistorical and Cognitive Analyses of Exploration

How can we sort out these discussions of how people relate to robots, which have become entwined in perhaps even more confusing discussions of how science relates to exploration? To ground this discussion, two very different analyses are required: a *sociohistorical* perspective on exploration as a competitive, nationalistic endeavor and a *cognitive* perspective on exploration as an activity that is part of human learning and has a refined character in scientific work. It is this second, cognitive interpretation that is involved in the claim that all field science involves exploration and that people are exploring Mars using robots. It's important to realize that the MER scientists, too, are caught up in the sociohistorical implications of their mission, as they see and reflect on the implications of leaving their tracks on another planet. Both the sociohistorical and cognitive perspectives are legitimate and useful for understanding how MER has changed field science and the experience of being a planetary scientist.

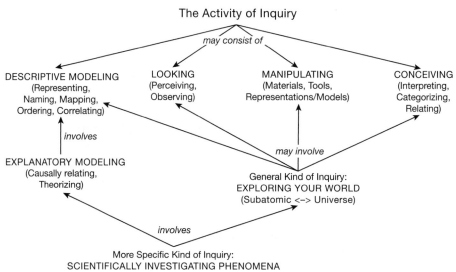

FIGURE 3.6
Aspects of the human activity of inquiry: the interactive, cognitive processes of looking, manipulating, conceiving, and descriptive modeling may occur when exploring. Scientifically investigating involves exploring, but also involves explanatory modeling by causally relating and theorizing about phenomena, as well as *systematically* observing, using tools, manipulating materials, correlating data, and so on.

In a broad study of "Exploration and Science," ranging from Humboldt to robotic space exploration, Reidy, Kroll, and Conway describe scientific exploration as always being a "multilayered social, cultural, political, and economic drama."[68] My objective here is to add another layer—actually a different dimension of analysis—namely, cognitive methods and processes.

From the sociohistorical perspective, Stephen Pyne analyzes "exploration" in terms of its organizational form and cultural purposes: "Exploration as an institution is an invention of particular societies.... It derives much of its power because it bonds geographic travel to cultural movements, because it taps into deep rivalries, and because its narrative conveys a moral message."[69] For example, proponents of human spaceflight such as Robert Zubrin have referred to space exploration not only as having scientific purposes, but also as tapping into an aspect of human nature and constituting a privilege to shape the destiny of the human species: "Why do it? First for the knowledge.... Second, for the challenge. People thrive on challenge and whither without it.... Third, for our future ... [Mars] is our New World.... We have the opportunity to be the founders ... of a new and dynamic branch of the human family."[70] The historical metaphors are clear here; they are related well (and provocatively) in Pyne's exposition.

In contrast with such broad sociohistorical characterizations of human endeavor, a cognitive analysis considers the character of the mental and physical work being accomplished (figure 3.6). From a cognitive perspective, the most general activity may be called *inquiry*. In saying that inquiry is an activity, we mean that it is something people do—it occurs as behavior in some place, over time. Inquiry involves fundamental forms of perceptual-motor coordination: looking (angling your head, moving your body), manipulating (turning over an object, scraping an object, adjusting a hand lens, applying a filter to an image in a photograph editing program), conceiving (objectifying a perceived feature, relating sizes and locations), and modeling (naming, making a map, listing related features and correlating their properties). As we may glean from field notebooks (figure 3.2), drawing and looking are intricately related, enhancing observation of detail and relationships. Analyzing intellectual processes as interactive, mental-physical behaviors was advocated by John Dewey and may be contrasted with other theories such as behaviorism (which downplays the active constructions occurring in the brain) and cognitivism (which downplays the active constructions occurring in the world).[71]

Inquiry occurs in many forms in the everyday world; for example, physicians, financial auditors, the police, congressional committees, and parents attending to their children's affairs engage in inquiry, each with their own practices. For our purposes, we can distinguish two kinds of inquiry: the activities of *exploring* and *scientifically investigating*. Historians and social scientists characterizing Earth and space studies[72] have found it useful to distinguish, for example, the interests and accomplishments of Magellan (an imperialist-explorer) from Darwin (a scientist). Exploring is the more general activity, and it occurs in a wide variety of social settings. A young child crawling through a room is exploring. An entrepreneur may explore a potential market for a new product idea. To explore is simply "to search through or into: make a first or preliminary study of" some phenomenon or condition. Although some would argue that the race to the South Pole was merely grandstanding, how to get there and return safely involved inquiry about the place—determining the nature of the ice and weather patterns was essential.

Scientific investigation involves exploring in this broad sense, but it adds a more disciplined, systematic aspect to the modeling process. Notice that "modeling" is something people are doing all the time; to model is to objectify a situation, thing, process, and so on by articulating its properties and often how they are changed over time or through actions. Thus, a child touching a lightbulb and saying "hot" is modeling the world. Scientific modeling involves going beyond isolated descriptions and simple if/then relations to form systems that are often formally described in some special notation or language, including classifications, causal models relating structures and processes over time, and more general theories about "how things work," which become the disciplinary principles of meteorology, geology, biology, astronomy, and so on.

In planetary science, exploring is pervasive; it is an activity using tools such as robotic probes, telescopes, lunar rovers, and so on and framed by a scientific investigation. That is, we are not simply exploring to see what is out there and making maps. Instead, scientific questions and theories are driving the identification of targets of inquiry (e.g., the suspected oceans of Europa) and the invention of new instruments for gathering data (e.g., including a drill with the Mars Science Laboratory).

Both aspects of the MER team's inquiry—exploring a world and scientifically investigating phenomena—are illustrated by Squyres's remarks at the August 18, 2004, press conference regarding their encounter with the rock formation Clovis in the Columbia Hills:

> We have now found one of the best examples of bedrock the rovers have seen.... One of the things that we have been looking for in the Columbia Hills since the very start has been signs of what a geologist would call "alteration." Now alteration is a very broad term; it simply means changes to the chemistry of a rock. And there are lots of things that can cause alteration. One of them, and the one that intrigues us, is water, and so we've been searching for signs of alteration by water in the Columbia Hills ...
> we're beginning to think that we may have found it. And we're working out what the implications of that are, but it's starting to look very interesting up in the hills.[73]

The scientists were not just moving through the hills; they were exploring—looking and manipulating materials with named conceptions in mind (e.g., "bedrock") and generally surveying the terrain. They were actively searching for signs of alteration, regardless of the cause. Identifying Clovis as a feature of interest was part of modeling the terrain: breaking the morass of rock and soil into geologically conceived areas and objects, then classifying, naming, and relating these. Preliminary investigative study of the rock outcrop—including ratting of Clovis and using the MI, the APXS, and the Mössbauer, then relying on theoretical signatures of atomic and molecular structure (i.e., models of how the instruments behave when exposed to different minerals and surfaces)—revealed that water may have altered this rock.

In general, a "discovery" is the finding and/or identification of a phenomenon (or its absence). For example, not finding the quantity and distribution of carbonates on the martian surface that some scientists had expected is one of the discoveries of the past decade.[74] The discovery of water alteration of bedrock in the Columbia Hills occurred after a deliberate attempt to find that phenomenon. A discovery may also occur serendipitously when exploring without scientifically intending to examine something. For example, Spirit's dragging wheel turned up subsurface white material, a discovery made by the team when routinely examining photographs. Analysis using Mini-TES led to the further discovery that this material was nearly pure silica.[75] "Discovery" may also be used to name a form of inquiry in which we explore with the intent of finding new objects or phenomena in some setting (e.g., Cook's attempt to discover a south-polar continent). But it is clear enough and simpler to say that one is exploring.

In summary, scientific inquiry is a specific kind of exploring activity. However, it doesn't necessarily involve exploration in a geographic sense of broadly characterizing an unknown territory. A scientific discovery is a discovery that occurs within scientific inquiry. During MER, this inquiry focuses on the morphology and chemical composition of rocks and soils; the general shape and layout of landforms, the geographic aspect, can be broadly discerned from orbit (figure 3.4). Like Humboldt, the MER scientists are scientific explorers.

The cognitive analysis of inquiry (figure 3.6) emphasizes the fundamental nature of *agency*, a pivotal concept for comparing the relation of people to today's robotic systems. In particular, the coordination of perception and action involved in looking and manipulating is necessary for being a physical agent. MER has some of these capabilities. A person is also a cognitive agent—people interact with their environment while they are adaptively modifying their goals, behaviors, and interpretations. In human cognition, perception and action are *conceptually coordinated*. In simple terms, a person has an ability to blend ways of thinking about a situation and improvise behaviors while acting—indeed, the essence of having an experience, as Dewey emphasized, is reflecting and interpreting as we behave, whether speaking, writing, drawing, singing, or dancing.[76] By combining the cognitive abilities of people and the physical capabilities of MER, field science on Mars can be conducted from Earth.

Today, no machine is capable of dynamically coordinating looking, manipulating, and interpreting in the manner of a person; that is, we do not know how to build a robotic geologist.[77] But because "exploration" in a geographic sense can be useful without ongoing cognitive interpretation and analysis, we can imagine building in the next few decades robotic reconnaissance and prospecting systems that work relatively independently of people. The implications for designing future robots go beyond the scope of this book; figure 3.6 sketches some of the capabilities that need to be brought together.

The cognitive perspective—scientific investigation is a kind of inquiry that combines scientific modeling with exploration—is pivotal for understanding how the scientists relate to the rover and for talking in general about the relation of people and robots. Instead of a superficial story of "robotic geology," we are left with an actual puzzle: how does the integration of instruments, software tools, and mission operations more generally in MER's design enable the scientists to engage in such an inquiry while remaining on Earth?

To recap, in this chapter we have teased apart two perspectives for talk about "exploring" and "being explorers"—a *cognitive analysis* that considers the work the scientists are doing and how they accomplish it (the notion of inquiry) and a *sociohistorical analysis* that combines a historical and often romantic way of seeing space exploration through past exploits on Earth and imagined adventures of the future.

We have also seen that the contentiousness (and confusion) about "science" and "exploration" in the press and space community is occurring in a sociopolitical sphere and that these terms are interpreted quite differently in a cognitive analysis. In particular, the term *exploration* in the name "Mars Exploration Rover" refers broadly to the exploring inherent in field science. The scientists are making discoveries by using the rovers as robotic laboratories in a way that combines traditional fieldwork and laboratory work. The discoveries and exploration are about *physical entities* (i.e., having to do with landforms and morphologies, such as depositional layers) as well *conceptual entities* (i.e., the ontological distinctions such as "alteration" and causal relations constituting theoretical models of geochemistry, meteorology, and so on).

The MER scientists understand and experience, and sometimes blend, both the cognitive and sociohistorical perspectives in how they conduct their work and talk about the mission's importance. We can see from the nature of field science why the scientists might view themselves as *traveling on Mars*. Sometimes this cognitive experience, which we will see has an essential role in their investigation, is accompanied by an emotional feeling of the historical place of the mission, which they conceive as "being explorers."

Both the scientific and romantic themes are worth pursuing, but—like the scientists'— my interest anchors in the technical part, namely understanding how MER's design enabled high-quality field science. Fundamentally, how is the experience of being on Mars possible? In the next chapter, I analyze the rover and accompanying organizational roles, schedules, and ground software as an exploration system designed for field science. How people, software, instruments, and processes come together is quite complex, and though thoughtfully justified in the mission's design, their synergy in practice goes beyond even what experienced scientists and engineers anticipated.

4

A New Kind of Scientific Exploration System

After reading about MER over the years, it may become easy to take the story for granted and summarize it simplistically: "Rovers are exploring Mars; scientists are learning about ancient aqueous processes." But when Steve Squyres says in a lecture, "We could have done much more if we were there,"[1] a host of questions arises: what was MER lacking? Is the solution a technology we can afford to build soon? Or is the limitation inherent in the very concept of doing field science with a robotic laboratory? Or is Squyres's belief, generally shared by the scientists, just an early impression that will go away after they learn to live with a new way of thinking and working? Answering these questions requires understanding the MER operations concept and the psychology of working remotely in a bit more detail. The topics we consider here include the nature of an exploration system, how MER was teleoperated, the psychological experience of *presence*, and the capability of commanding MER every day (one-sol turnaround). Together, they reveal MER as a system—a combination of people, technology, and operations—that enables scientists to engage in Mars field science without leaving their home planet.

The MER Exploration System

In general, by "exploration system" we mean a particular kind of *work system*—a combination of organizations ("staffing"), vehicles, tools (hardware and software), facilities (buildings, habitats), procedures, schedules, and so on—that specifically enables human spaceflight and/or planetary science missions. Whether robotic systems are involved and the relative roles of people and robotic systems may vary; each configuration constitutes an exploration system. Thus we can describe the exploration system of the Apollo lunar landings, the International Space Station (ISS) assembly missions, the Hubble spacecraft under normal operations, the Hubble repair missions, the planetary probes (see table 2.1, for example), and so on. One may also construe the development, training, debriefing processes, and perhaps even job or career trajectories during a long mission like Cassini as being part of the work system, because these factors all contribute to the ultimate quality of the work. But by a work system, we usually mean the configuration during the "operations" phase of the project.

Our present concern is the MER exploration system as designed for carrying out the work of planetary field science on Mars (MER's "surface operations"). Regardless of whether field science is performed by people in person or mediated by robotic operations, the essential capabilities of the work system include: observation methods (e.g., sensing instruments, cameras), mobility methods (i.e., a means of moving through the landscape and placing instruments), and manipulation methods (e.g., digging, scraping tools). Because the MER scientists are working remotely, software tools for driving and controlling the rover play a central role in the operations. Broadly speaking, I will refer to this software capability as "virtual reality" (VR), but I significantly refine this concept as we go along.

Because a robotic laboratory is a major community and national undertaking, the MER scientific work is not controlled by a small group of investigators, but is rather managed as a formal organization tied firmly to various scientific, university, and government organizations. Consequently, the MER work system is designed for enabling teamwork and must specifically support a *communal* manner of doing scientific work. Further, the style of fieldwork shifts from the traditional small-group, opportunistic use of tools to deliberative and systematic targeting and timing, as illustrated by the scientists learning how to use the RAT at El Capitan. Finally, given the importance of the public's support, as well as the strong inherent interest previously evidenced during the Pathfinder mission, the MER exploration system was designed to facilitate public awareness, through press releases and a regularly updated website with status reports, personal stories, raw Hazcam and Pancam photographs, and maps.

The system of MER surface operations processes and tools for the engineering and science teams was developed iteratively through trial and error in tests over three years and refined during the mission itself. The parallels with Pathfinder operations—involving a rover, two teams, use of VR, and so on—are mostly superficial correspondences. In particular, the VR and planning tools used in MER in 2004 are considerably refined from the late 1990s software that related photographs, science plans, and rover commands. Notably, a tool for bridging science plans and engineering "sequences" was not defined until October 2003, after a tedious, error-prone task was identified during operational readiness tests over the preceding six months (figure 7.3). Effectively linking the work of the science and engineering teams was particularly crucial in MER because daily commanding based on the previous sol's results—managing the programs, memory, and power required for the RAT, MB, APXS, Mini-TES, Pancam, and MI, compared to Sojourner's single APXS instrument—was a pivotal mission operations requirement.

The daily commanding requirement is based on the nature of field science; however, its necessity was not obvious to everyone. Indeed, the preliminary design review (PDR) raised serious questions about nearly every aspect of the MER work system, including daily commanding, living on Mars time, the planning and sequencing software, and how the organization would operate.[2] To understand why some proposed aspects of MER's surface operations were contentious and why getting the whole system together was not easy, we need to consider the different ways a robotic system can be operated.

Telepresence through Synergistic Telerobotics

Robots can be defined as computer-controlled machines. The robots we are using today and for at least the next few decades are not like the humanoid, conscious beings found in science fiction. As Launius and McCurdy put it, robots are "mechanical servants under human supervision, not independent machines."[3] People are always "in the loop"—people program robots to carry out certain procedures that specify behaviors under changing conditions. How often and easily people can change the programs during the mission significantly changes the nature of robotic operations, both in the capabilities of the robotic system and the capabilities of the people to participate in what the robot is doing.

A *telerobot* (also "teleoperated robot") is a robot that is remotely programmed; that is, the robot's program of operations is created by people and transmitted from a distance during routine operations.[4] All modern spacecraft and most satellites are telerobotic. Most people do not realize (and as an academic computer scientist, I was amazed to learn) that spacecraft operations software is commonly written, tested, and "uploaded" to the satellite during its "cruise phase," the period between launch and the start of the prime mission in orbit, flyby, or on a surface. New routines may be validated during the mission, changing the behavior of the robotic system. For example, MER's program for navigating over long distances, altering its path slightly to avoid obstacles, was modified during the mission.

Using telerobotics to do field science on a planetary surface was first performed on the moon by the Soviets during the Lunokhod mission during the early 1970s (table 2.1); controlling it required careful planning, and operations were direct, rather than automated in a comprehensive daily program.[5] Recall that Surveyor did not enable field science because it did not enable scientists to explore the landscape—it was not a rover. Scientists interested in doing fieldwork on Mars have discussed the possibilities of using telerobotics for some time. Stoker characterized this kind of surface exploration system as "a new paradigm of exploration where humans and robots operate synergistically to achieve exploration goals.... An alternative to providing mobility to take astronauts to scientific field sites would be to provide them with teleoperated robots that could be placed virtually anywhere on Mars."[6]

Telescience by remotely reprogramming scientific instruments in an adaptive manner as the investigation proceeds was introduced in Viking operations.[7] Today, we take for granted that a spacecraft needs to be reprogrammed as knowledge gained during the mission reveals new opportunities and interests; plus, we've come to expect that a one-sol turnaround is routinely possible (see table 2.1). But inventing a spacecraft operating system that could accept new command sequences and designing mission operation tools and processes so the ground team could respond adaptively were significant innovations in the 1970s.

In adaptive mission operations, as the process was first termed on Viking, the *turnaround time* is how long it takes to create, validate, and transmit new commands ("uplink") after receiving and interpreting the results of the previously programmed operations ("downlink"). In terms of round-trip transmission of data and commands, the minimum turnaround time is bounded by the distance of the robot from the operations team, varying from a few seconds for the moon to forty minutes or more for Mars to hours for Jupiter and beyond. In practice, it takes time for people to interpret what is happening and decide what to do next. The minimum time is important only for direct teleoperations, such as physically adjusting instruments (e.g., aiming a camera, digging a trench); this mode is precluded at distances much beyond the moon. In addition, more complex instrument operations involving imagery scans, spectral analysis, chemical processing, heating, and so on may inherently require minutes to hours (e.g., the APXS and the Mini-TES). Instruments have been scheduled for much longer continuous automated operations (e.g., Viking's organic analysis required 12 days[8]).

The "synergy" suggested by Stoker involves telescience in which scientists are interacting in tempo with the actual time remote operations require. The effect is that one uploads a command, receives feedback as soon as the operation is complete (e.g., for a movement/manipulation, feedback indicates what motion occurred; for instrument data, the results of the scan or analysis), and has the opportunity to command the laboratory again, as soon as the previous results have been interpreted and a new plan constructed.

For Viking, the turnaround time ranged from seventeen to twenty-one days (figure 4.1). The mapping to martian sols would necessarily vary, but published reports about operations refer only to days. Indeed, according to Viking Project Manager Jim Martin, an attempt to experimentally work on Mars time as an eight-day training exercise was deemed too physiologically problematic, so the team worked on Earth time.[9] Nevertheless, Viking's operations on Mars were organized by and described in terms of sols.[10]

Activity/Day	1	2	3	4	5	6	7	8	9	10	11	12	13	14	15	16	17	18	19	20	21	22	23	24	25	26	27
VIKING																											
Science Plan 1	▒	▒																									
Engineering 1			═	═	═	═	═	═	═	═	═	═	═	═	═	═											
Sol Uplink 1																	■		■		■						
Science Plan 2							▒	▒																			
Engineering 2									═	═	═	═	═	═	═	═	═	═	═	═	═						
Sol Uplink 2																							■		■		■
SOJOURNER/MER																											
Science Plan 1		▒																									
Engineering 1		═																									
Sol Uplink 1	■																										
Science Plan 2		▒																									
Engineering 2		═																									
Sol Uplink 2	■																										
PHOENIX																											
Science Plan 1	▒	▒																									
Engineering 1		═																									
Sol Uplink 1	■																										
Science Plan 2		▒	▒																								
Engineering 2		═																									
Sol Uplink 2		■																									

FIGURE 4.1
Comparison of science commanding turnaround on Mars missions.

Each science planning cycle of Viking (two are shown, numbered 1 and 2) nominally involved two days to develop a conceptual plan for three science uplinks to occur over a six-day period; two weeks were required for generating the programs. This cycle restarted every six days during the forty-three sols of high activity on Viking 1 (i.e., seven cycles) until commanding was reduced to accommodate the arrival of Viking 2. Daily communication relays occurred during this period, allowing for receipt of science data and engineering operations.[11] Sojourner pioneered routine daily science commanding, as used on MER, with a full cycle occurring in one day (the "tactical" planning process occurs after the rover completes the sol N program and continues during the rover's night in time for an uplink on the morning of sol $N+1$). Phoenix is similar, except that more advance planning is possible for a stationary laboratory; a "strategic science plan" was created on sol $N-2$, revised to a "tactical science plan" on sol $N-1$, and uplinked for sol N operations.

Despite nominally requiring two weeks to complete the uplink program, urgent changes could be realized in two sols, a strategy that allowed responding to "next-to-last" downlink data[12] (e.g., when a problem retracting Viking 1's boom was detected on sol 2, a command sequence to test a diagnosis was prepared for sol 5), and repeating an already validated sequence could be done overnight.[13] Note that the results of planning cycle 1 became available only near the end of engineering cycle 2 (during "command and sequencing"); thus, "late changes were virtually inevitable."[14] For MER, the turnaround objective for routine operations was one sol. That is, on the Viking missions, the scientists would wait more than two weeks after making a nominal plan to get results; for MER, the effect was operating the rover every day, enabling a feeling of synergistic operation—indeed, of being there on the planet.

A one-sol turnaround is all the more striking when you realize that the work requires getting the previous sol's data to the scientists, interpreting the state of the rover and the previous sol's operations, and discussing what to do next. The MER engineers then had about twelve hours to convert the scientists' plan to a verified sequence that could be uplinked. This accomplishment was made possible by having engineers working surface operations twenty-four hours a day in three shifts; by having the scientists provide the engineers with "skeletal activity plans" that anticipated the overall form of the next sol's activities; and because the scientists were able to make further data-gathering decisions on the basis of rudimentary daily analyses, informed by ongoing, more thorough analyses that individuals and small groups were pursuing in parallel and presented in the science assessment meetings.[15] Furthermore, as explained in "Being a Textbook Scientist" in chapter 7, the scientific focus was on gathering data and feedback about the quality of the data, less on testing hypotheses and shifting plans based on experimental outcomes.

The daily turnaround on MER enabled an experience of *telepresence*, "the sense of being at a real location other than where one actually is":[16]

Telepresence is a high-fidelity form of remote control which projects the sense of the human operator into a robot at a distant work site. Telepresence represents the marriage of two important technologies: virtual reality and advanced robotics. Using telepresence, a robot can be operated in such a way to give the user a strong sense of presence in a remote environment…. Human operation can be mixed with low-level control of automated functions.[17]

In MER, this sensing and display technology is not coordinated in real time—that is, through telerobotic programming—which enables projecting oneself into the rover's current state, imagining future actions, and creating plans (programs) that carry out the desired operations. Through this combination, the scientists experience *efficacy*: being an agent on Mars and perceiving and acting in the martian environment.

Telepresence has been experienced in undersea exploration (e.g., using the Jason robotic vehicle[18]), surgery, process control, and military applications.[19] The experience of "presence" is a philosophical topic that has been transformed by the creation of tools emphasizing visual feedback and enabling telerobotic operation. The academic research community studying presence—including psychologists, interface designers, and social scientists—is considering how it is possible for people to operate devices remotely. Various studies compare being physically present to mediated operations (e.g., using computerized maps and photographs), also called "virtual presence."[20] Anticipating the importance of telepresence in missions like MER, anthropologists in the 1990s studied the experience of presence in field geology "to develop specifications for domain-based planetary exploration systems utilizing virtual presence."[21] Recently, some JPL engineers working on MER have renamed the visualization methods used since the Voyager mission virtual presence in space (VPS) technology.[22]

Because the MER scientists experience being on Mars, they are not speaking meta-phorically when they describe themselves as explorers—not merely because they are exploring the nature of Mars conceptually, but also because they experience being travel-ers on Mars as people doing field science. Thus the MER exploration system contradicts a prediction by Pyne that planetary robots will be the explorers, not people: "The chief novelty unveiled by space travel will be the character of exploration itself, that the explorer may be—ought to be—robotic and virtual."[23] MER demonstrates that the virtual presence provided by field science telerobotics with daily commanding gives people the experience of being the explorer. The robotic component moves mentally into the background—for example, the rover's Pancam movements are wielded transparently by imagining standing somewhat above the rover and turning one's head.[24]

A plausible perceptual-motor experience of seeing and moving on Mars is possible because MER is a semi-anthropomorphic telerobot (figures 2.2 and 2.3)—it was built to mimic aspects of the human form, with stereo vision "eyes" (a bit lower than was first desired), an arm that can place instruments against surfaces, and wheels that enable aiming (targeting), moving on long traverses, and an impressive (though not reliable and never fast) climbing ability. This design is not just a convention of how we'd like a robot to appear; rather, it greatly facilitates the experience of virtual presence. The embodiment of the scientists and the engineers into the local martian topography is essential for operating the robotic laboratory—people must adopt the perspective of being present on Mars as explorers to move and use MER effectively. This conclusion is somewhat unexpected and starkly contrasts with most of the space program poetry touting "intrepid explorers."

Now, to be clear, telerobotic laboratories could be operated on Mars in different ways. Stoker imagined the combination of two methods: automated (programmed in advance) reconnaissance prospecting the landscape combined with real-time direct control from a Mars base.[25] The Viking, Sojourner, and MER missions use a third alternative—"batch-operated" (cf. table 2.1) teleoperation without either automated surveying or direct manipulation of devices. By shortening the turnaround time from weeks to overnight and using a form of VR extending over months and years of operation, telepresence is greatly enhanced from the adaptive operations pioneered on the Viking mission and prototyped on the Sojourner mission. How this work system design was developed during the MER mission and justified scientifically merits further consideration.

The Development of Daily Commanding in MER Surface Operations

Although the natural advantages of the experience of "being there" while doing field science might now seem obvious, designing the tools, organization, and scheduling in the MER exploration system (called "The MER Mission System" at JPL) was not obvious or easy. The surface operations process for doing the scientific work and tending the rover included—for each MER—a three-shift team, dedicated meeting rooms and offices at JPL (later distributed), a suite of integrated computer tools, and the Deep Space Network and Odyssey satellite orbiting Mars for relaying communications.

Strikingly, Gentry Lee, Viking science analysis and mission planning director, cautioned the MER scientists, engineers, and managers during the preliminary design review in April 2001 that MER's surface operations posed challenges unlike any previous mission: "Of all the phases of the MER mission, the only one that is completely new is the surface operations phase. Entry, descent, and landing (EDL) is difficult, to be sure, but the EDL phase on MER is not significantly different from Pathfinder. Cruise is similar to dozens of other missions. *Surface operations is brand new*."[26] These remarks are intriguing because one might have thought (and to some degree the JPL engineers believed before the PDR) that MER's surface operations organization, tools, and schedules could be simply scaled up from Pathfinder/Sojourner designs. Gentry Lee said that this was particularly not true for the software for operating the rover: "In the software area I sensed the greatest amount of Pathfinder paradigm syndrome. It is simply not sufficient to take the Pathfinder ground software and scale it up slightly for MER. MER is an extended surface operations mission with much higher expectations for success."[27]

Living on Mars time for so long was new. Sojourner was commanded daily for eighty-three sols, reverting to two shifts living on Earth time after a month. Yet in saying "extended," Lee was thinking in terms of only a three-month mission—not multiple years! The complexities of MER's nominal mission were unprecedented and somewhat alarmed this senior Viking engineer: operating two rovers on different sides of the planet simultaneously (Viking operations were designed to put the first lander on a more automated mode before the second spacecraft landed), guaranteeing a nominal daily commanding cycle (Viking operations were designed to allow three-day turnaround only for last-minute changes), and coordinating three hundred people on three shifts living on Mars time for ninety sols—the initial, "prime" phase of the mission. As on the Viking mission, Lee was playing the "role of interlocutor," in which he adopted the task of ensuring "that the operational people and the scientists understood each other's needs and limitations."[28]

Although the Pathfinder mission had far more sophisticated automation than Viking (both on Mars and for ground operations on Earth) and included a rover, the mission was nevertheless less complex than MER in many ways: Sojourner was not as big and much simpler, with fewer instruments; the operations duration was much shorter with fewer mission objectives (reverting to two shifts on Earth time after thirty sols); and daily planning involved far fewer people using a simpler network of software. Yet analogies between Pathfinder/Sojourner and MER provided useful guidance for at least the initial design of the MER mission system.

Andy Mishkin played a key role in defining the surface operations and tools on MER as the mission operations system development manager, applying his experience as principal designer of Sojourner's mission operations. The initial descriptions of MER "ground software" refer to the MOS (mission operations system), a work system that comprises teams following procedures using GDS (ground data system) tools. But surface operations also include the science process proper, such as different kinds of scientist meetings, the science theme groups, and the SOWG.

Mishkin's discussion of Sojourner refers to the Experiment Operations Working Group (EOWG), a progenitor of the SOWG.[29] After a one-hour science assessment meeting to share information among disciplines, the instrument teams developed a joint plan. During the subsequent EOWG meeting, the engineer on duty detailed what was done the day before (the operations and what data was received) and spacecraft constraints for the coming day (e.g., downlink capacity and power). Scientists proposed what to do the next day and received immediate feedback from engineers on what would work and what wouldn't.[30]

MER's science theme groups were prefigured by Viking's shift from a team of PIs to egalitarian science teams (e.g., biology, radio science, physical properties) that included engineers:[31] "NASA space science to date has been organized around principal investigators, individual scientists who propose a specific study, develop an instrument for flight, analyze data received, and publish results. Viking instead makes use of individual scientific teams. Each team has a leader who represents the team to the Viking project, but the scientists in each team participate on an equal basis."[32] Building on such ideas, MER's surface operations evolved between 2001 and 2004 and even during the mission and were eventually refined into something "brand new," as Lee advised.

To understand the transition in thinking from Viking and Pathfinder to MER, consider that the history of Sojourner told by Mishkin is distinctly engineering-centric. Befitting what was essentially an engineering demonstration mission, he mentions the scientists and what they were doing only in passing. In contrast with the integrated science—engineering field trials for MER, Sojourner's "surface operations mode test" focused on the basic capabilities of rover-lander communications, not the scientific work. Stoker characterizes the ORTs as "mostly engineering" tests that either failed or were not well conceived (e.g., at first scientists could not print images). Unsurprisingly, given the limited opportunity for mutual learning, Mishkin mentions that during the mission scientists' requests "forced sequences to be many times longer than predicted during design phase."[33] Stoker says that all through the mission, she felt like she did not know how to use the tools—how to make things work.[34]

In the story of Sojourner, "the real work of engineering the rover" is not described from the perspective of the scientists study of Mars—enabling virtual presence—but rather from the perspective of managing the science integration team to stay within the limited weight, size, and power allowances of the tiny rover. Just getting an operational rover on the surface was difficult enough. With such a tiny rover, of the chosen scientific instruments, only the APXS could fly—very far from the notion of an anthropomorphic laboratory.[35] Nevertheless, according to Golombek, scientists and engineers worked closely together as a team during the mission's EOWG meetings: "There was never an us vs. them mentality."[36] Stoker emphasizes that JPL was "very responsive to suggestions and worked hard to implement solutions."[37] This positive experience probably contributed to co-inventing the much more complicated planning processes required for MER.

Originally, the MER science process was characterized in terms of coarse rover "operational processes" involved in reaching a target to which the IDD would be applied: "Panorama Day, Traverse Day, Approach Day, Contact Science Day."[38] The engineers focused on defining typical scenarios that the MOS would need to support. Over time, this high-level design was refined as the practical perspective of using the rover within a scientific inquiry led the scientists and engineers to define more specific operations that tended to combine the use of instruments: Pancam plus Mini-TES; RAT plus MI; drive plus image Navcam tracks. Rather than moving a few meters and targeting instruments on the most obvious rocks as Sojourner did, MER would involve traverses, spotting distant features, and planning on multiple time scales. Justifying and designing this daily commanding process required the scientists to articulate their vision for Mars field science more clearly to the review panels. Specifically, the scientists and engineers as a team had to invent practical methods for discussing and formalizing plans that could in less than a day become new computer programs, uploaded and allowed to run unattended on Mars.

Justifying Daily Commanding for Field Science

As we have seen, traveling virtually on a rover—using its tools and opportunistically planning targets and/or a path every day in campaigns lasting months or years—made it possible to use MER to do field science on Mars, distinguishing it from any previous Mars mission. Again, by "field science" we mean functionally replicating the practice of Earth scientists in scientifically exploring a place—not merely standing in one place and sampling what you can reach or being confined to your backyard.

The system required for operating the MER laboratory involved more than just "batch operations" or what JPL engineering called "remote commanding"—a linear operation of formalizing the command sequence overnight, transmitting the program after the rover awakes, and receiving data at the end of the sol. Rather, the MER exploration system needed to be a total system, a closed loop unifying commanding and data with science planning and supporting multiple-sol operations. For example, to enable continuity with a daily turnaround time, shared online databases were required for defining target names and associated photographs and instrument data.[39]

The incremental, honing-in nature of planetary fieldwork is explicit in the Athena proposal: "Key rock and soil targets will be selected on the basis of Pancam and Mini-TES data, and will be approached for close-up imaging and acquisition of Mössbauer and Alpha Particle X-Ray Spectra."[40] But more important, the entire mission was planned to be opportunistic and contingent on what was learned from day to day:

> A primary example of these are what we call "target of opportunity" activities performed with instruments on the IDD. On panorama, approach, and drive sols, work with the IDD is not the primary objective of the sol. However, front Hazcam images obtained from the rover's start-of-sol position—obtained, for example, at the end of the previous sol— can document the appearance of the work volume of the IDD. There will always be some Martian surface materials within reach of the IDD; these constitute targets of opportunity. If the rover's configuration permits an IDD deployment, one or more targets of opportunity can be investigated with IDD instruments in addition to the main objectives of the sol.[41]

In contrast with Sojourner, MER's schedule and organizational roles were designed for phases of "strategic" (longer-term) and "tactical" (daily) planning. Both the engineers and the scientists had defined subgroups of people doing these tasks. But why were daily operations proposed for MER seven days a week for the anticipated life of the rover? Why not follow a two-week plan more like that of the Viking mission? The primary reason given in early 2001 was to maximize science data return during an expected lifetime of 90–120 sols, during when dust buildup and cold could gradually reduce the number and type of operations possible.[42] For example, if dust buildup reduced the daily power available by too much, driving longer distances to reach new targets might become impossible; then the available areas to study and accomplish goals in using the instruments might become too limited.

Driving distance and feature variety were essential because MER was designed for an actual field science investigation with traverses to new areas, continuously revealing new surface features to study. In this respect, Sojourner was a demonstration of the daily movement, targeting, and instrument placement cycle (using APXS) that MER would put into practice on a more ambitious scale, combining multiple instruments in a variety of scenarios. The work would be more meaningful and would continuously extend into new areas; therefore, the one-sol turnaround time would allow *contingent planning*. Early design documents highlighted these aspects of the daily "uplink process":

- "Rover must interact with its surroundings, which change every time the rover moves."[43]
- Contingent planning makes operations highly dependent on stereo images with a display for visualizing terrain and interactively designating instrument targets (computed from geometric terrain models).[44]
- Given the state of the rover and results of previous sol activities, the MER team must generate and validate the next sol's uplink within about fifteen hours (originally thought to be nineteen hours, but could be as short as eight hours depending on communications relay availability).

The anticipated daily movement of the rover, either driving or moving the IDD to place or orient instruments, would entail waiting until the receipt of end-of-sol telemetry be-cause the position of the arm, the instruments, and the rover itself after programmed movement is nondeterministic; that is, the rover's state cannot be reliably predicted suf-ficiently precisely. In particular, slippage of the rover or failures would lead to a different state—including location, data, and instrument operation—than had been expected from the previous sol's program (notice that a three-sol plan was tractable on Viking [figure 4.1] because it didn't rove). Reasoning from the perspective of the rover, for new operations to be provided for each sol—and given that the rover was solar-powered and must therefore do most of its operations during the day—the team concluded that "command turnaround must generally occur over a single Mars night."[45] This constraint and conclusion was not new; the Sojourner rover team also needed telemetry from the previous sol to generate a plan: "After long traverses, Sojourner might end up a yard or more from its target"; also, the autonomous drive operation would halt prematurely from being too cautious or from erroneous sensor readings.[46]

Also, unlike prior missions, MER's strategic plans outlined operations for each sol over the entire forthcoming two weeks. Sojourner was operated mostly from day to day, and Viking's "long-range plans" covered operations that would not *start* until at least two weeks in the future. Eventually, MER operations also required creating and repeatedly adapting much longer-range plans for "campaigns," the lengthy traverses and multiple-month studies of complex locations (e.g., Endurance, Home Plate), with even longer-term plans for handling the martian winter, Mars-Earth solar conjunction, and the multiyear traverse to Endeavour.

But MER's PDR team in 2001 didn't anticipate campaign planning or even the first solar conjunction that would occur nine months after landing. Instead, the "brand-new" aspect that worried Gentry Lee involved the number of instruments and the simultaneous operation of two rovers. Operating MER goes well beyond the methods that worked for Sojourner because of the variety of potentially interacting alternative actions on a given sol—move the arm or rover, measure in situ, remote sensing, communications—all of which must be completely specified as targets, parameter settings, and durations and then sequenced (ordered) within power, thermal, timing, memory, and bandwidth con-straints. Because failures are possible that make subsequent operations irrelevant or even damage the rover, each daily program was designed so that a failure would halt the planned sequence and carry out default actions such as taking photographs. Furthermore, each sol's uplink included a "runout" sequence (not involving traverse or contact) in case the rover didn't receive the next sol's commands on schedule.[47]

The plan for daily commanding while living on Mars time presented at the March 2001 PDR was still being articulated in the months before the review meeting.[48] Certainly, the beneficial implications that would emerge were not all realized initially. Much more salient at first was how the proposed overall dual-rover process differed from any previous orbital or lander mission. For example, the focus of the Athena proposal was on instru-ment operations—using the payload to achieve science objectives—not the scientists' mental and personal engagement in the inquiry (though Squyres and colleagues suggested that the overarching slow pace would encourage wider participation).[49] One of the PDR reviewers was particularly unclear regarding how to think about the relation of strategic (long-term) and tactical (daily) planning:

> The relationship between the strategic and tactical planning activities for surface opera-tions was not clear from the presentations. Based upon the discussion, it appears that there is not a consistent view among all Project personnel either. Specifically, it is not clear how much the strategic plan will define the daily activities. Without further specification, the possibilities range from a Galileo/Cassini approach, where the skeleton "strategic" plan maps out the specific objectives of each day well in advance, to a Mars Pathfinder–like mode, where each day or two is planned from scratch with an overall strategic "plan" as a general objective or guideline. More likely, the desired mode is somewhere in between. This is a fundamental philosophical issue and could drive many specifics of the system design. Therefore, this process should be specified clearly and unambiguously (and soon). One way of specifying it would be to define the specific products that come from the strategic planning process and how they are used in the tactical process.[50]

What is missing here is a concept of how field science proceeds—in particular, the pacing MER would both allow (by daily commanding) and deny to the scientists (by progressing so slowly through the vast terrain). One response to the reviewers was the invention of the "sol tree," which reified a mental image of a field science plan created by the long-term planning (LTP) group. The plan looked ahead about two weeks and enumerated alternative operations as far as could be seen (figure 4.2). As the reviewer implied, the plan maps out the work well in advance as on Galileo/Cassini, combining and filling in reusable "templates," but these skeletal plans and enumerations would be adapted daily as for the Pathfinder mission. For a lander such as Viking or Phoenix, the reachable terrain can be mapped once and used throughout the mission (incorporating new trenches and instrument targets over time). When MER traveled, the terrain had to be mapped for each new location; in addition, multiple locations were often visited with their features named and specific targets for instruments pinpointed in a single week. The ability to create and display maps and photographs on a computer screen changed tremendously during these thirty years (figures 6.5, 6.8, and 8.7), but because these tools were prototyped by the scientists and not the JPL engineers, they had been used only rudimentarily for Sojourner.

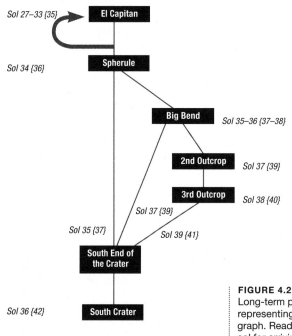

FIGURE 4.2
Long-term plan for Opportunity on sols 27-42, representing the "soltree" created on M26 as a graph. Reading from top to bottom, the earliest sol for arriving at a given site is indicated, with latest planned sol in brackets {}, if all alternative routes are taken. For example, sol 35–36 {37–38} means "as early as sol 35 and 36 (two sols at this site); no later than sols 37 and 38." The loop at El Capitan indicates repeated work at that site, ending by sol 35. Compare to figure 6.8.

Even someone familiar with field science on Earth, knowing that it is carried out by individuals or small groups, would wonder how the science team could be expected to make so many decisions about very different instruments every sol for months. The PDR reviewer appropriately commented, "The number of science people that are planned to be involved in making daily decisions in a quick turnaround fashion seems unrealistic"—in other words, there appeared to be *too many* scientists involved. Another reviewer concluded, "This scenario for surface ops is extremely optimistic. Relying on daily command loads and an overnight sequencing process should be re-evaluated. Recommend looking at extending sequence loads to at least two to three days, if possible. The current approach will surely exhaust the teams doing the work and will increase the risk of making mistakes." [51]

Asked to go back and reassess or better articulate the operations plan, the team determined that the original plan was best, summarized by Squyres in the "Assessment Regarding 'Frequency of Uplink'" presented at the Critical Design Review in December 2001: [52]

- Operating on Mars time is the best way to assure that the rover will be commanded every sol.
- Commands should be based on the previous day's operations (not just "runout").
- The science team must have adequate participation in all operational decisions.

Also responding to the PDR recommendations, Mishkin articulated the tactical surface process (daily commanding turnaround process) and staffing approach in subsequent presentations. [53] The reality, of course, is that after several years of conceptual development, practical tool design, and experimental operations in ORTs, everything more or less worked out fine. Heeding the PDR reviewers' concerns, the science and engineering teams pulled together staffing, scheduling, processes, tools, and training to accomplish every sol commanding.

How Daily Commanding Enhanced Scientific Quality

Commanding MER every sol was originally viewed as necessary to maximize the amount of scientific data gathered and exercise the rover's capabilities within the expected short lifetime. But a fundamental serendipitous benefit emerged: the intensive daily activity, with sufficient feedback to assess the previous day's plan and move on, significantly increased the scientists' engagement with their rover's activities, making the rover's state and what might be done next something they thought about every day, month after month, and ultimately year after year.

Staying engaged and projecting themselves into the landscape fits the scientists' experience of being in the field—a turnaround of more than two weeks, as on the Viking mission, would instead require additional effort to remember where they were or what they were doing. Even with distributed operations, more than a year after the nominal mission had completed, Jim Bell, the lead Pancam scientist, said that daily turnaround "helps keep us all excited about the exploration aspect of these missions: many times, we don't know what we're going to be doing that day with a rover until we see the results from the day before."[54] This routine daily process, operating a wide variety of instruments on a mobile vehicle, with people distributed across the world, continuing for years—inconceivable during Viking's time and only prototyped for Pathfinder/Sojourner—had been made possible by the advances in constraint-based planning and scheduling programs, shared databases, and web-based graphical browsers developed over the preceding decade and refined in training as well as during the mission.

In short, daily commanding arguably improved the quality of the scientists' work, further increasing their productivity in selecting suitable targets (figure 4.3). Long-range planning, in which looking weeks out would often mean scanning for or anticipating features not in the immediate landscape, was therefore potentially more scientifically conceptual and strategic than on the Viking mission.

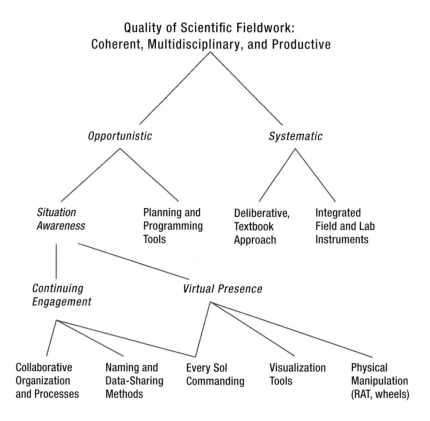

Quality of Scientific Fieldwork:
Coherent, Multidisciplinary, and Productive

Daily commanding also facilitated balancing the disciplinary interests of the scientists. With the handful of instruments available and the competing interests of scanning the atmosphere, doing detailed geochemistry analyses, and moving on in a broad geological survey, being able to trade off opportunities every few days enabled a cooperative process and enhanced the experience of working together as a multidisciplinary team. Consequently, the daily command turnaround helped implement Squyres's vision that the MER team would not just cooperate in sharing a common platform, as in other missions, but would also collaborate in sharing a common inquiry—namely, studying Mars holistically. Further discussion of this distinguishing characteristic of MER appears in chapter 7 under "One Instrument, One Team."

Being opportunistic is enabled by knowing what could be done (situation awareness) and timely execution through commanding tools. Awareness in turn includes longer-term engagement (so individuals are participating) as well as oriented projection into the immediate setting to engage field skills (virtual presence). Many more details could be included in figure 4.3, such as having multiple communication opportunities during the sol, daily retargeted imagery, visual alignment of images and measurements with multiple perspectives, documentation of observations, and integrated software tools.

In conclusion, maximizing productivity, planning as field scientists, maintaining awareness and engagement (virtual presence), and making the use of the rover a coherent single scientific investigation were all enabled or enhanced by daily commanding (figure 4.3). This is an important lesson learned because aside from productivity, none of these benefits were listed in the formal presentations to justify daily commanding during the MER review meetings.

FIGURE 4.3
Relation of requirement for coherent, multi-disciplinary, and maximally productive field science to *features* of the MER work system (in italics), supported by the methods employed.

5

The Mission Scientists

Scientists joining NASA from academic positions in areas of computer science, psychology, or the social sciences, such as anthropology, soon learn that "scientist" has an unfamiliar meaning at NASA. *Scientists* are those who work in the fields of inquiry that drive space science projects, especially geology, physics, chemistry, biology, and astronomy. Other members of the team are called *engineers* (e.g., computer scientists) or *human factors specialists* (e.g., social scientists). Their research topics are not treated in missions as being of scientific interest in themselves; the planetary scientists' interest in having such people around is practical—their work is instrumental to successful operations and is highly valued. This distinction of roles, which can seem jarring at first, is inherent and essential for organizing missions, in what is after all a choreography of hundreds of PhDs from different disciplines. So when working at JPL, the human-centered computing team just smiled and weathered our odd separation from the rank of "scientists" when participating on the MER mission. We sat in the back and observed, making design suggestions in science operations review meetings and privately to managers, which was both our role and privilege.

Some years later, in April 2009, I met Steve Dick, who was then leaving a productive tenure as head of the NASA Headquarters History Division. After explaining a bit about my MER writing project and then my background as a cognitive scientist, his quizzical expression suddenly resolved into a broad smile: "Ah, so you are a scientist!" Oh, no. The implications were unmistakable: NASA mission histories were the turf of historians—and I had just revealed myself to be a scientist writing about other scientists, providing fodder for the official reports that would actually go into NASA's archives. After a few rounds of this, you stop worrying what you are called as long as people are happy to have you participate.

This chapter is about scientists' identity and specifically how roles are formally defined in the MER organization and informally constructed by members of the science team, as they—like the engineers, human factors researchers, and historians—find a niche and develop what is often a blended identity in a joint enterprise.

The Athena Team and Collaborating Scientists

Broadly speaking, the MER scientists are planetary scientists—professional scientists studying the solar system in a wide range of disciplines, including astronomy, astrophysics, geophysics, geochemistry, mineralogy, earth science, geology, and biology. Table 5.1 characterizes the MER Athena Team (see figure 5.1/plate 6), the group of scientists who formally led the mission. The three subgroups are the Athena proposal team (as constituted for the Mars Exploration Program in 2000), the twenty-eight "participating scientists" selected competitively from eighty-four proposals by NASA in spring 2002 before launch of the spacecraft,[1] and eight scientists selected to replace departing team members in fall 2005.[2] Most of the non-Americans are associated with MER instruments developed in other countries and were consequently part of the original proposal. University and government agency affiliations dominate and are balanced. In addition, the February 2004 science roster lists eighty-five "collaborators" (scientists associated with specific science team members, most from the original Athena team) and fourteen students.

FIGURE 5.1/PLATE 6
Athena Science Team and other MER team
members in early 2003. *Image credit:* NASA/
JPL. *Source:* JPL MER Mission, "Team Photos,"
http://marsrovers.jpl.nasa.gov/people/photo1-
br.html (accessed February 17, 2004).

TABLE 5.1

MER (Athena) Science Team membership categorized by country and affiliation (university, government
agency, corporation, and nonprofit organization) and when the scientist was invited to join the team

MEMBERSHIP PHASE	SUBTOTAL	COUNTRY		AFFILIATION			
		USA	OTHER	UNIV.	GOVT.	CORP.	NONPROFIT
ATHENA PROPOSAL (2000)	22	17	5	12	8	2	0
PARTICIPATING SCIENTISTS (2002)	28	26	2	11	13	1	3
SECOND-ROUND ADDITIONS (2005)	8	8	0	3	4	1	0
TOTAL	58	51	7	26	25	4	3

Note
The data summarized here as well as the institutional totals come from the cited 2002 and 2005 press releases and a
spreadsheet provided by John Callas on February 17, 2004.

The 157 people directly participating in MER science operations (Athena team, collaborators, and students) are from 37 different institutions, of which 17 institutions have one representative, including 4 of the 8 scientists chosen in the second round. The distribution suggests a large number of research project and collegial relationships, dominated by Arizona State University (19 team members, including Phil Christensen, Mini-TES lead), JPL (16), Cornell (15; Steve Squyres, PI, and Jim Bell, Pancam lead), USGS (14; Ken Herkenhoff, MI lead), Washington University (11; Ray Arvidson, Deputy Principal Investigator), and NASA (11).

To better interpret observations of the MER science team in 2004 and put their work in the broader context that has developed since then, I interviewed seven MER scientists,[3] chosen because they were at different points in their career and participated in the mission in different ways (table 5.2). I had met Michael Sims, Jim Rice, and Nathalie Cabrol during Haughton–Mars analog expeditions in the Canadian Arctic in 1998–1999, and we had spoken many times over the years; plus, Sims and Cabrol worked at NASA Ames. Dave Des Marais is a key long-term planner who also worked at NASA Ames; we were introduced in Pasadena in February 2004 while I was observing MER operations. Michael Carr and R. Aileen Yingst were suggested when I asked for a list of scientists who were on the extremes of retiring and just joining the planetary science community. Carr's tenure as SOWG chair overlapped my time in Pasadena, so I was eager to hear his thoughts about the events occurring then; he also lived in a nearby community, so an in-person interview was possible. I was briefly introduced to Yingst through Rice at an Ames meeting of MER scientists in July 2006 and recalled their conversation about Home Plate. Both Yingst and

TABLE 5.2

Background and role of the interviewed MER scientists (age is at the time of the landing of Spirit in January 2004)

SCIENTIST, AGE	PROFESSION	PHD FIELD, YEAR (UNIVERSITY)	WHEN JOINED	SCIENCE TEAM ROLE
MICHAEL CARR, 68	ASTROGEOLOGIST (RETIRED)	GEOLOGY, 1960 (YALE)	CO-INVESTIGATOR	SOWG CHAIR
DAVE DES MARAIS, 56	BIOGEOCHEMIST	GEOCHEMISTRY, 1974 (INDIANA)	CO-INVESTIGATOR	LTP
MICHAEL SIMS, 54	ARTIFICIAL INTELLIGENCE AND ROBOTICS	COMPUTER SCIENCE, 1990 (RUTGERS)	CO-INVESTIGATOR	PDL & LTP
STEVE SQUYRES, 47	PLANETARY SCIENCE	GEOLOGY/PLANETARY SCIENCE, 1981 (CORNELL)	PRINCIPAL INVESTIGATOR	SCIENCE LEAD
JIM RICE, 44	ASTROGEOLOGIST	GEOMORPHOLOGY 1997 (ARIZONA STATE)	PARTICIPATING SCIENTIST	GEOLOGY & LTP
NATHALIE CABROL, 40	PLANETARY GEOLOGIST	PLANETARY GEOLOGY/ SCIENCES, 1991 (SORBONNE)	PARTICIPATING SCIENTIST	GEOLOGY
R. AILEEN YINGST, 34	PLANETARY GEOLOGIST	GEOLOGICAL SCIENCES 1998 (BROWN)	2ND PHASE PARTICIPATING SCIENTIST	GEOLOGY

Rice were interviewed by phone (Yingst was doing fieldwork in Oregon); the other interviews were in person.

After completing an article based on these initial interviews, it became clear that the material enabled and required a longer exposition. The frequent mention of Squyres by the scientists and his role suggested that I interview him also. Further, I realized that the working relation to the engineers was a key part of the MER science story, so I inquired about engineers who played a central role and might provide a different angle on science operations. I selected Jake Matijevic because of his significant role throughout the mission and Pathfinder experience; in 2006 he was also chief engineer for MER. I selected Ashitey Trebi-Ollennu because he was instrumental in developing key rover hardware; like Cabrol, he was educated outside the United States; and in 2006, he was the most experienced rover planner. Squyres and Matijevic were interviewed in person at JPL; Trebi-Ollennu was unavailable then, so he was interviewed later by phone.

Given that more than 150 scientists participated in the MER operations during the first few years of operations alone, the choice of the seven scientists is not claimed to be representative of each of the possible dimensions for classifying the group (age, discipline, mission experience, affiliation, country, role, etc.). Instead, the sampling is *across* these dimensions, emphasizing age and hence different career origins and experience. This choice seems suitable given the focus on how working with a rover changes the nature of field science. The interviews reflect how previous background affects personal experience, especially with respect to expectations (how you want to study Mars) and motivations (why participating in this mission is important to you). A story about the *development* of the MER mission, as opposed to the scientists' experience during the mission, would necessarily include many other JPL engineers and managers, as well as the developers of the Mössbauer Spectrometer, the Mini-TES, and so on.[4]

Even focusing on science operations, people might be selected based on the science theme groups (e.g., geochemistry and atmosphere scientists), the international instrument teams, laboratory versus field experience prior to the mission, collaborators versus Athena scientists proper, or the broader JPL organization (e.g., most notably Mark Adler, Spirit Mission Manager and originator of the concept of adapting the Pathfinder crash-bag scheme for landing the Athena rover instrument package,[5] and Joy Crisp, the project scientist, whose role is described subsequently). Even one such additional dimension would easily double the number of interviews and could begin to constitute a scientific survey.

In contrast, the present study is an ethnographic story that seeks to provide insights about how the MER scientists relate to the rovers, the public, and each other. It would have been my pleasure to spend years talking to more people and transcribing and writing about all their thoughts and memories, which would constitute studying the team, rather than the essential characteristics of the fieldwork. I believe the excerpts provided here are sufficient to illustrate a range of perspectives that reveal both the practical and personal aspects of using a robotic laboratory to do field science on Mars.[6]

Referring now to the information in table 5.2, which lists the interviewed scientists, "co-investigators" refers to the authors of the accepted Athena proposal. The indicated profession is in response to the question, "What kind of scientist are you?" Reflecting in part the long gestation of the mission, a scientist's age correlates with the phase of participation: Athena proposal co-investigator, MER project participating scientist, and second-phase participating scientist.

The MER scientists I interviewed epitomize the blending of scientific identities. Strikingly, most received a PhD in a specific field and later combined and applied their skills to planetary science. By the 1980s, some universities were giving specialized degrees in planetary geology. The three younger scientists (Cabrol, Rice, and Yingst) knew early in life that they wanted to work in the space program. Squyres transitioned from being a geology undergraduate to planetary science as a graduate student, when he was invited by Carl Sagan to participate on the Voyager team. The older two scientists (Carr and Des Marais) were drawn into the Apollo lunar program through pivotal meetings and projects. Sims became involved through an artificial intelligence (AI) organization at NASA, working on robotics and operations software.

Planetary science by one account was created by virtue of the existence of the lunar samples returned from Apollo and the process of defining and managing planetary missions; in particular, the Planetary Science Institute was formed in 1972. Sims explained that before Apollo, "People were paid as astronomers, but not as planetary scientists." In effect, NASA created the discipline to frame the work and to develop a cadre of participants: "They created a domain and an infrastructure—astrobiology is a more modern variant—and they created a community."

As a rule, MER scientists view participating on this mission as the culmination of their careers, regardless of their age. Being a member of the MER science team is the realization of their professional identity. However, most of a MER scientist's career is spent in fieldwork, teaching, and/or mission planning, not in science operations. Subsequent sections elaborate and illustrate the progressive blending of identities that has enabled to them to do fieldwork through a rover, their shared interest in exploration that a martian expedition realizes, and how the team's objective of "figuring out Mars" requires transcending a disciplinary focus. Problems faced by the outliers by age, Carr and Yingst, reveal the challenges in getting on board the rover team. Participating is made difficult for one person (the now retired member) by the long experience of working differently and for the other (the later joining member) by the team's tacit knowledge from years of operating the rovers.

To properly frame these personal aspects, I begin with a more formal view from the top of the science team organization by considering the responsibilities of the project scientist and the principal investigator.

The Project Scientist and Principal Investigator Roles

Planetary science missions have been organized in different ways for developing, packaging, and using the instrument "payload." Responsibility is allocated among a JPL project scientist, usually multiple principal investigators (PIs), and science teams specialized by instrument and/or discipline. Use of these roles in missions may vary. For example, in MER from 2004 to 2006, Squyres was the Athena PI and Joy Crisp was the MER project scientist, but for the Mars Science Laboratory, the project scientist, Grotzinger, combines both responsibilities with the help of two deputies (one of whom is Crisp). Understanding MER science operations, which means understanding what accounts for the quality of the work, requires considering both the project scientist and PI positions.

Viewed from within the MER science team and following the scientific details of the mission either in public or in the press, an observer might see only Squyres, the sole PI and visible, articulate spokesman, along with the science theme groups. Crisp's work occurred outside the science team's meetings, but is pivotal for understanding what "working with a robotic laboratory" entails in practice from an institutional/NASA perspective.

The relation of the PI and project scientist roles is not obvious for academics; university projects are managed by a PI responsible for the scientific integrity and productivity, often assisted by co-investigators. A research project may be managed by co-PIs who allocate the conceptual, logistic, and presentational responsibilities. A project like MER is so large that two mindsets are required—one (the PI) looking down, in a sense, to manage the details of the ongoing work, and the other (the project scientist) looking up to relate and represent the scientists' perspective to other activities within the MER project, including the scientific community, engineering teams, management, and public affairs. Simply put, one could say that Squyres as PI led the (Athena) science investigative team in their inquiry, while Crisp as project scientist was the scientist lead for NASA's MER project.

Crisp, a volcanologist who was the lead scientist for Pathfinder's APXS, served as coordinator and representative for the MER scientists and the scientific process, attending to the broader perspectives necessary for the scientific success of the project.[7] She reported to the project manager and had a line of communication with NASA headquarters; Squyres reported to Crisp. The remaining scientists and their organizational bodies, such as the SOWG and science theme groups, reported to Squyres. In this science management hierarchy, Crisp's responsibilities included ensuring that the team had necessary equipment and support, plus seeking ways to maximize (within project constraints) the science return. The project scientist makes scientific judgments for the project; for example, prior to a public announcement, Crisp arranged a peer review of the science team's finding that there was once flowing liquid water on the surface of Mars. Crisp interacted with engineers for assessing risk versus science return for critical decisions, such as driving into Endurance and handling instrument problems. She gave approximately 150 interviews and 100 outreach talks from 2004 through 2006; she approved all NASA/JPL press releases about MER. In some of these activities, the MER project scientist works with the MER science manager, who handles more administrative duties (e.g., science facilities). Starting in 2000, the science manager was John Callas, to whom the PI also has a dotted-line relation.

Squyres's formal title is Athena principal investigator (not "MER PI"); thus, just as Athena is conceived as a payload wrapped in the engineering wherewithal for delivering, landing, and operations on Mars, the PI and scientists' work is wrapped within the JPL organization and management process. Squyres's *organizational* responsibility is therefore to "ensure that the investigation's definition, science payload design and development, planning and operational support, and data analyses are successfully accomplished within the Project schedule and the team's available resources." Strikingly, his project job description emphasizes the science *payload* rather than, for example, ensuring that the team is developing scientific theories about Mars.

In thinking about conventional field science, one might not expect a geologist to be charged with "successful hardware integration and implementation of the experiment," including supporting calibration of the hazard cameras and navigation cameras. Of course, a wide range of university research PIs have such instrument design and development responsibility, such as laboratory psychologists, chemists, physicists, biologists, and astronomers. Still, an observer of the El Capitan operations in February 2004, unaware of the early conceptions and failed proposals and the preparations leading to launch, would be surprised perhaps at the combination of scientific and engineering knowledge required of the Athena PI.

Although I had studied MER science operations for several years, I first realized that Squyres was more than the leader of the science team when I opened *Roving Mars* in 2006 and discovered that the phrase "science team" is not mentioned until page 230. More than half of the book is about the rovers, the remainder about operations. Details about thermal vacuum testing, fuses, cables, and batteries are not what geologists usually write about. Squyres explains that his involvement as PI in the components of the rover, their interactions, and testing enabled most of the other scientists to "focus on their real job, that is, doing the science." It was necessary for the scientists to be familiar with the instruments and the way that they work before using data from them, but attending to preparing instruments for launch would have just been a distraction.

So what does the PI need to know about engineering and what was he doing when the rover was being developed? He replies, accenting words in a manner reminiscent of Carl Sagan:

> What I tried to do during the development phase as PI was to sit back, *understand the whole system well enough* that I could see the parts of the system that were most going to influence the quality of the science. For example, the layout of the solar arrays affects how much power we have, which affects how long the vehicle's going to last, which affects whether or not we can do the science that we want to do. The size of the solar arrays was hugely important! We're still going on sol 1400 because Randy Lindemann figured out

how to pack great big solar arrays into a tiny little volume, then people like Kobie Boykins actually figured out mechanically how to pull it off. That was hugely important. And so I spent an enormous, inordinate amount of time looking at solar array issues because I *knew* that was going to be important for science. I could list a hundred things for you. What I would do is try to find the high leverage items.

Crucially, from the start in conceiving instruments and rover proposals, Squyres saw his job not merely as a scientist in determining where to go on Mars, what measurements to take, and how to interpret the data, but as a *science systems engineer*, someone who designed and tested the rover holistically, connecting parts and processes to ensure the quality of the science:

I saw my job during development as kind of science systems engineering—to be able to look at the total system and find the parts of the system whose quality one way or another has the greatest bearing, the greatest influence on the science. Because what would happen is that the engineers would each be given part of the problem to solve, and they would go off and solve their part of the problem. But they didn't maybe have the broad overview of what we were trying to do scientifically. They had a specific job, a set of requirements to meet, to build a bracket, build a widget, build a wheel, build a computer, build a rover that would meet those requirements. But, they might not have enough insight into the science to realize that even though this one part of the rover meets the requirement, if you could make it just 5 percent better or 10 percent better it would have a huge influence on the quality of the science. And there were some things that were tremendously important and others that were much less so.

Another example of high leverage is the speed of image processing performed onboard the rovers:

We didn't do a good job when we first started defining the requirements of talking about how *fast* the software for doing image processing onboard the rovers had to be, and initially it was painfully slow. If you wanted to take a series of Pancam images, you could only take one image every six minutes. We got that down to thirty or forty-five seconds. If you want to take these great big panoramas and you have only so many minutes during the day, that's an order of magnitude difference in how much data you can acquire. And so I screamed bloody murder about that one forever.

In his role as science systems engineer, Squyres had responsibility for twelve different payload elements—six instruments on each rover—which he shepherded through the design, test, verification, delivery, and integration processes.

Strikingly, with Squyres as the sole PI of Athena, the design of the science instruments and operations could be under his control—the job description says, "wholly and solely accountable for the Athena science payload instruments"—a freedom and responsibility offered by the mission's initial formulation within NASA's Discovery program:[8]

Our mission was structured very differently from most. The norm on planetary missions, Cassini, MSL [Mars Science Laboratory] has been that you have individual PIs for every single instrument. And then they compete with one another for resources. You have a project science group that is chaired by the Project Scientist. And they have the job of herding these cats—it's more like herding saber-toothed tigers, I think—getting everybody to play nice together, to work together. The situation on M–E–R *(enunciated)* was fundamentally different in that we had a single PI. And so I had responsibility for the entire science payload.

The consequences for the quality of the overall design are sharp and fundamental:

> It meant that I could optimize the performance of the total science payload. It was a science systems engineering task, where you have this very complicated multidimensional problem. But I had a lot more levers that I could push as PI for the whole payload, than a typical PI does for a single instrument. So, for example, if I had one instrument that's doing well and one that's in trouble, say financially, I can move resources from one to another. I can push money around, I can push people around, I can do the things that I need to do to solve one problem. It's a system optimization problem, and the higher the level at which you do the systems optimization, the more parameters you have at your disposal to play with.

Crucially, Squyres's oversight and control focused on the integration of the instruments— so rather than worrying about how independent PIs worked together to produce a coherent result, he could directly focus on designing the instruments to work together:

> When it came to the *design* of the payload, rather than have a bunch of PIs each go off, and they each design their instrument and then they propose it, and you've got to figure out some way to make it work together, I was able—at the *outset*—to design the microscopic imager, the RAT, the Mössbauer, the APXS, so they had fields of view that fit together nicely so that everything worked when you went to look at a given spot on the martian surface. Just design the whole system so the pieces fit together. And that was nice!

This idea of integrated instruments has deep implications throughout the mission. It means not only that the instruments align in useful ways, but it extends to the choice of instruments, so the rover as a whole is designed to remotely carry out the scientists' work. This holistic perspective relates to the concept of the "rover as geologist surrogate" and has ripples throughout the design and experience of the mission: (1) the scientists are organized according to disciplines not instrument payloads; (2) the scientists as individuals can project themselves into the rover's body to orient themselves to the work setting and imagine what they would do next; (3) the metaphor of the "robotic geologist" shaped stories about the rover and created a kind of romance about their journey, which reinforced the science group's identity. These ideas are developed further in subsequent chapters as we consider the experience of "being the rover" (chapter 6), the "communal scientist" (chapter 7), and the "personal scientist" (chapter 9).

A perhaps farther-reaching lesson is learned by considering how Squyres was able to accomplish this science systems engineering role and whether it can be repeated. His unusual background in conceiving Athena and developing the instruments through a series of proposals provided important knowledge that he was able to exploit as PI. He knew the people, the instruments, and the science that needed to be done, and he put these pieces of the puzzle into place. Squyres stresses that "it was an unusual situation, it's hard to replicate on an MSL or another mission because [MER] is structured in a very different sort of fashion." Having eight or even a dozen PIs in the science group poses a different set of organizational problems that a "small class," capped-budget Discovery mission avoids.

Before elaborating further on how the systems engineering of a coherent rover gets mapped into the concept of a coherent team, I would like to focus on the backgrounds and motivations of the individual scientists. Their ability to blend identities and their shared strong desire to be scientific explorers helps explain how a top-down organizational design was possible.

Progressive Blending of Identities

A modern scientific career involves inventing and adapting different lenses for viewing the world and being effective, as increasingly complex technologies inherently require interdisciplinary perspectives and problem-solving methods. The blending of science and engineering expertise developed by Squyres over the years enabled him to become a "science systems engineer." Indeed, the very design of MER as an integrated rover, which Squyres conceived, requires that the members of the science team draw from their own experiences and capabilities a blend of interdisciplinary and scientist-engineer thinking to meet the needs of the mission (figure 5.2).

Accordingly, the scientists describe themselves as a blending of identities, analytic ways of thinking, and tools they have acquired in their career. Ultimately, the scientists distinguish themselves and thus rationalize their participation in the mission by forging special combinations of several topics and tools. However, as explained in the next few sections, although they share many broad experiences about exploration and being on mission teams, within the project they have different interests and care about different aspects of the work.

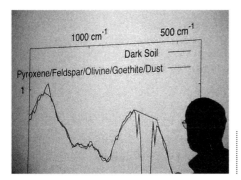

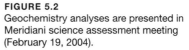

FIGURE 5.2
Geochemistry analyses are presented in Meridiani science assessment meeting (February 19, 2004).

To begin, consider the eldest member of the team, Michael Carr. He was brought to the US Geological Survey (USGS) to work on lunar geology by Gene Shoemaker, viewed by many as the originator of astrogeology. In the early 1970s, Carr became involved in Mariner (a Mars orbiter) and then developing the cameras for the Viking orbiter and the image-processing system at JPL. Asked if he was always interested in engineering, he said, "No, not really, you know, things came along. Geology was always the basic theme." All of the scientists similarly referred to their academic roots to describe their interest and capability. Asked if his interest is more broadly exploring space, he said, "I think it's mostly the geology. Mars is the geologist's planet. And if it weren't for Mars, I'm not sure I'd be an astrogeologist. I've been involved with other missions—Galileo, to Io and Europa, and the moon. But for me, they don't have the fascination that Mars has, where every different kind of geologic process has been at work. So I really do think it's not so much the space science."

Dave Des Marais is eleven years younger than Carr, so in doing a geochemistry PhD, he had the opportunity to work on lunar rocks: "I started out my formal education in chemistry and then got interested in geology. My dissertation for my PhD was on lunar samples looking at carbon and nitrogen, elements of biological interest in lunar samples. So I guess that means I was doing planetary science at that point. What a time to be in grad school if you're going to get involved with NASA stuff! All the Apollo stuff is happening, and the rocks are coming back." Asked if he had a particular interest in the space program in the 1960s, he said, "No. Actually, quite frankly, I might have been off working for some chemical company if it weren't for becoming interested in going into caves as a hobby when I was an undergraduate. I just happened to stumble into a dorm room with a guy who was with an outing club and he started talking about these caves."

Des Marais arrived at NASA in 1976: "The year of Viking. The rationale for my hire was interpretation of results from Viking and pursuing ideas it engendered." So now Des Marais's professional identity concatenated further: "Biogeochemistry is the phrase that I like because I started out life in chemistry and then I got into geology and then I grafted biology on top of that. So—chemistry, geochemistry, biogeochemistry. *(laughs)* It follows my life story. There's a journal by that name too: *Biogeochemistry*." Yet Des Marais is grounded in chemistry:

> One thing that ties together a lot of what I've done in my research is the chemistry of carbon. Because I studied carbon in lunar rocks, which is demonstrably a nonbiological theme, I began studying carbon in volcanic rocks and how carbon is processed by the Earth, deep in the mantle, and so forth. More broadly, this is the carbon cycle—how is carbon, the element, processed by planet Earth and how is it processed by the biosphere, and what are the interactions there? Those are relevant questions as we go to Mars.

But when NASA asks for job role, he writes "space scientist."

Steve Squyres is a bit younger than Des Marais and was an adolescent during the 1960s. His interest in the space program was strong, but he didn't view himself as having a space exploration career:

CHAPTER 5

I always followed the space program very closely when I was a kid growing up, I think more so than most kids my age did. I remember John Glenn's flight vividly. I was six at the time. I was glued to the television through that whole flight. So it was always something that I was interested in, but I didn't view space exploration as I was growing up as a viable career option. It was something people on television did—some other kind of person. I was just a kid from New Jersey.

In college, Squyres realized he could identify with "people who explore Mars":

I didn't really start to look at space exploration as something that I might do until I got to college. And then it was my junior year in college that I took a course on the results of the Viking mission that I started to realize that, you know, I'm in the same room with people who explore Mars and they don't walk with their feet six inches off the floor. They're just normal people, too. And so, maybe I could do this…. But I didn't start to think of it seriously until I was in college.

Squyres was then fortunate to be invited to work with Carl Sagan:

I was told that Carl Sagan wanted to see me. I had never met Carl. And so I go by his office—he had a secretary to kind of guard his office; it was hard to get in to see him. I went in and sat down and there he was, Carl Sagan, big as life. And he says to me he'd like me to be his grad student for the Voyager mission. And my little heart starts going pitter-pat, because the Voyager science team was just the original people who had been selected for that team and there were two card-carrying geoscientists on that team at the time, Hal Masursky and Larry Soderblom, both from USGS. Because a government laboratory doesn't have graduate students, I had no way of getting involved with Voyager as a geoscientist. And then, Carl Sagan wants me to be his graduate student on the Voyager imaging team. So that was pretty exciting.

Starting with an interest in cameras, Squyres acquired engineering colleagues and knowledge that had not been part of his formal training, preparing himself for his eventual systems integration role:

I didn't initially have an engineering background. My training was as a geologist, and then as a grad student I was doing planetary science. My first introduction to engineering came when we first began to develop what was the precursor to Pancam. Back in the early 1990s I took a one-year sabbatical at Ball Aerospace when they developed the original Pancam. I knew nothing about engineering. So I went out and I just bought books, on electrical circuit design, and how motors worked, CCD [charge-coupled device] detector technology, and I read a lot. And we built this camera, this instrument. I decided I was going to learn everything about that.

Not many geologists take a sabbatical in an aerospace company. One cannot but wonder whether the smooth collaboration between the scientists and engineers, required to an extreme on MER, was a reflection of Squyres's initiative and interest in bridging the gap in knowledge and perspectives.

As PI, Squyres's identity has indeed broadened considerably:

> I did kind of hardcore research in planetary science for many years. Now what I do is this kind of bizarre combination of planetary exploration, robotics, and management—all in the service of geology on another world. The fun thing about my job is that it requires a whole bunch of different areas in which I need to do things; every day is different. So I've got scientific challenges, technical challenges, engineering challenges, management challenges … communications, politics, all of that factors into the job.

Asked how he might describe this profession to a young student today, he said, "A planetary explorer. I don't know." The blending of roles and expertise can become so complete that the scientists lose track of academic categories and cannot find one that fits. So Squyres has moved from geology to planetary science (by degree) to being a planetary explorer—though of course such a label ignores the extent to which he is often an engineer.

Jim Rice, three years younger than Squyres, was also a teenager of the Apollo era; from the age of seven he grew up wanting to be involved in the space program. Rather than viewing space exploration as "something people on television do," as Squyres said, he directly identified with the heroes: "I wanted to be an astronaut like almost everybody." He's now obviously pleased to participate in a series of Mars missions.

Still a teenager when Carr was managing the Viking Orbiter, Rice was strongly influenced by that mission and identified with the camaraderie of the team and being an explorer:

> I remember as a kid, thirty years ago this summer, I was seventeen years old, watching the Viking landings on Mars. I followed it as much as I could back in those days—there was no Internet. I remember thinking, "It'd be great to be on a team, to be one of the geologists like Hal Masursky, to work on that team, to be part of it, to see these pictures for the very first time, trying to figure out what's going on up there."

Rice says his current work could be described in different ways:

> There's a lot of things you could call it—space scientist, planetary scientist, planetary geologist. I like the term "astrogeologist" on my business card. When I told people when I was in graduate school that I was a planetary scientist, they'd said, "What? Sanitary scientist? What are you doing?" With astrogeology, they click in automatically. Because of the "astro" [prefix], they know pretty much what you are talking about.

Shoemaker (who had earlier brought Carr to California) set up a branch at USGS in Arizona, where Rice worked as an intern in the 1980s. After the Mars Polar Lander crashed on Mars in 1999, Rice was without a mission. He said, "For a while, I was thinking, how am I going to pay bills and eat? I had no job, my job went away when Polar Lander crashed." Rice's life revolves around missions: "I have met nine of the twelve astronauts that walked on the moon and always try to meet and talk to folks who worked on Apollo. I went to Robert Goddard's home and field where he launched the world's first liquid fuel rocket March 16, 1926, this past July. It was a pilgrimage of sorts. Anyway, I basically live, eat, and dream space!"

Nathalie Cabrol is four years younger than Rice; she also did her graduate work in the 1980s, focusing from the start on planetary science. Her career exemplifies a scientist-engineer identity, for her scientific work has often involved robots and other technology applied to the study of biological processes in extreme environments. Asked if she still views herself as a planetary geologist, having had a biology and robotics focus the past decade, she replied:

> I would say I am an explorer. Geology is one part of it. Definitely as you go along, you meet people, you see places, and your interests grow. I am not a biologist. I don't have the expertise. But I definitely developed an interest about extreme environments and extremophiles by exploring other places, trying to explain what on Mars could happen in the past. And definitely I have a curiosity for our planet as a whole, which still involves planetary geology, obviously. But I am interested in Earth more as a system, a system that needs a balance between the biosphere, hydrosphere, climate, and how these balances change with time fluctuation or crisis. I am interested in the exploration of this system.

Cabrol said that by virtue of working with robots, "My interests go way beyond geology at this point in time." Like others, she acknowledges she is still working in an area where she is most expert. "But definitely I am an explorer within my own exploration."

By working with robots, the nature of exploration itself becomes a topic of investigation; in particular, how do we explore hostile environments? Cabrol describes meeting a computer scientist working on robotics as a postdoctoral scholar at NASA: "That was the start of a ten-year relationship…. We have been continuously collaborating with each other, enriching each other with our experience, and nurturing each other." This collaboration shaped her scientific identity: "But I wanted not to stay only a geologist, I wanted to understand what were the advantages and constraints of robotic exploration, so I could become a geologist who could contribute to robotic exploration by knowing in what kind of frames and constraints I could work with, then the step beyond was really to make those systems better for exploration. Exploration strategy came about there." At the same time, the interest in geologic processes led to drawing analogies with Earth: "I wanted to understand, what could those lakes have been looking like on Mars? And this is what drove us to high altitudes and lakes on Mars. This is what we are exploring right now—trying to have various angles of the same thing."

R. Aileen Yingst, six years younger than Cabrol, is the youngest member of the MER science team, having received her degrees in the 1990s. For Yingst, being a member of the MER science team is rooted in her desire to be an explorer:

> I knew I had the bug, the mission bug, a long time ago. I have not taken the traditional path of getting a tenure position and then moving on into research. To go way back, I have always wanted to do space science—the exploratory aspect of studying other planets. I've often said that I *(laughs)* do space science because I couldn't join Star Fleet [from the *Star Trek* entertainment franchise]. Two hundred years ago, I might have been on the Lewis and Clark expedition. I really enjoy the discovery and exploratory aspect.

And to me, missions are the pinnacle of that vanguard, that exploratory, pushing the frontier attitude. So I have always wanted to do space, and since I realized I could be involved in missions, I've always wanted to do missions.

Michael Sims is actually just a few years younger than Des Marais; however, because he is a computer scientist, with different motivations, it's worthwhile to consider him separately. Sims illustrates yet another scientist-engineer blended identity, as he participates as one of three engineers in the MER science team—the "technology people" with a computer science and/or robotics background. Sims recognizes that some people on the team, like Des Marais, were not originally specialists in geology but became geologists "in order to do what's necessary in order to do the task." Sims has not made that conversion because geology is "not particularly interesting" to him. The aspects of the mission that relate to physics are interesting, and "the elements that relate to exploration are fascinating." By default, he falls into the "engineer" category for mission participants, but computer scientists do not usually refer to themselves as being engineers. So what's the right name for Sims's profession?

I struggle with this. What's the label you walk around within a team that has a single label ["scientists"] and you're supposed to fit inside, also with pretty much a single label? I cannot give a label, but I can tell you what I do. What I try to do is take technologies and make them work for the science we have at hand. I see my job as to figure out how to make the science work while using all the technologies we have in our bag, which are robotics and software, which are my primary degrees.

Thus Sims, like the others, grounds and justifies his participation by citing his academic degree: "I introduce myself as doing AI and robotics, that's my background. It's because I know robotics, I know what's going on, I know the software for the robotics."

One complexity in defining himself on the mission is that Sims works at NASA Ames, but JPL dominated his area of specialization during the MER mission: "For political reasons, almost all the robotics components are well-defined inside the JPL bubble. So there's not a huge arena for a contribution in that, but there is a place on the software side. So that's mostly what I've done—visualization software to make jobs easier."

Sims's work is always practical, sometimes narrowly defined by software problems, "doing software for software sake." He accepts the instrumental role of making jobs easier for others: "Sometimes that looks like sweeping the floors. That's not a problem." Here he shows the humility of a member of a tight team in which people disregard formal job descriptions to do what needs to get done for the organization to succeed.

Sims's thoughts fit Kohler's analyses of scientists who work in the interstices of disciplines or methods, especially between field and laboratory: "Practices are not just ways of making knowledge but also ways of life, and making a career has a social logic that may or may not coincide with professional credibility and output."[9] By attempting to be flexible and responsive in what Kohler calls "the border zone," Sims only sometimes realizes his identity as a computer scientist: "Sometimes it looks like I actually get to do things that make sense in terms of what I was educated to do." Although he is professionally focused on missions, having been a member of the Pathfinder team in 1997, his participation on the MER team is a compromise:

> It's fun. Is it satisfying career-wise? It's probably not. As a career, it's not especially a bright decision. In the sense of a traditional view of what a career is, or in terms of what my goals are, it's not necessarily consistent with most of my goals. But there is a single set of parameters that it is consistent with, which is that I'm all about exploration of space, and being part of the MER team is doing everything I can to make it work. So that's where I come from.

Perhaps correspondingly, Sims is the only one of the scientists interviewed who expressed a strong interest in being a PI of a future mission. Participating on MER is instrumental to this goal: "The reason for doing MER is that MER is good training to do the robotics we need to do for human exploration. That's a game that I can play and understand the rules of. Every new thing I take on with respect to space is filtered through the goal to get human exploration to the moon and Mars, and permanently there, in our lifetimes."

The Allure of Exploration

Despite the variation of scientific interests and capabilities, the MER scientists are united in their passion for exploration, for being explorers on Earth and Mars. They recognize, without diminishing their dedication to the scientific work, that the joyful excitement of being on Mars as part of the MER mission dominates their experience.

Carr tells us we are attracted to exploring Mars because its visible landscape combines mystery with a sense of possible understanding: "Mars is an alien planet yet not so alien as to be incomprehensible. The landscape is foreign yet we can still recognize familiar features such as volcanoes and river channels. We can transport ourselves through our surrogate rovers to a surface both strange and familiar and readily imagine some future explorers following in their paths."[10]

Yet exploring Mars taps something more general and powerful in our personal experience. As Rice relates, exploration can be exhilarating and intoxicating:

> When I was on Devon Island [in the Canadian Arctic], the first year that we went up there, I took off driving *(laughs)* on an ATV and just wanted to keep driving! I saw one butte on the horizon, I'd drive to that one, then I saw another one, of course there's always something, I kept going and going. The expedition leader was trying to rein me back in. He got so frustrated. He chewed me out when I got back there. He was yelling at me, he was blowing his horn, all this kind of stuff, "What were you doing?!" I said, "I don't know, I just had to keep going to see what was out there." That aspect of it really drives me, being one of the first, being the first to see something.

Exploration, described by Rice as "seeing something for the first time," is a common enthusiasm for the MER scientists. They describe their work with the excitement many people might associate with attending a sports event. For them, many days on the mission—particularly in the first few years—were like being on an adventurous trip. What is surprising is that the details of the scientific study then take on a secondary role. As many of the remarks by Rice, Cabrol, and Sims suggest, the geochemistry or geology or even software development story becomes a rationalization—a means of being part of the mission.

As explained in the discussion of cognitive versus sociohistorical perspectives of "exploration," being a virtual explorer for the scientists is grounded in scientific inquiry, so it refers to experience of traveling on the planet, discovering and analyzing soil and rocks, and so on. But this technical perspective of scientific discovery is imbued by the broader, sociohistorical meaning of the mission, in particular the emotional feeling of significance from being the first to be somewhere or to do something.

Rice explained how this broader motive can drive scientific work:

> I'll tell you, to be honest with you, the thing that interests me the most and gets me going all the time: it's just the sheer aspect of exploration and discovery. Just seeing something for the very first time—being the first person to put your eyeballs on that, or being the first person to notice that—the sense of wonder and awe of seeing something the very first time in history. I mean, we've seen things that no one in human history has ever seen, and I've seen them first! That to me, that's what gets me going, more than figuring out whether this rock has more olivine than that one. The detailed nitty-gritty science analysis is interesting, but if I had to say what drives me, motivates me, gets me really excited, it's just got to be exploration and discovery.

Rice describes being in the Antarctic, venturing into "places that no one had ever been before." He had been cautioned by others:

> Remember, be careful, no one's been out here, we don't know what we're going to find, be on your guard. That's intoxicating! It's this eerie feeling … I don't know, it's probably as close as I'm every going to get to going to another planet, personally, myself … I was able to go places in Antarctica that no human being had ever seen before, up to that point, the area where we were was very isolated, and not many expeditions had ever been there. I remember walking around the corner in this field, and the wind coming around there, and not knowing what's there, but you had to go and see, and keep going! It's an eerie kind of feeling, like you're not supposed to be here, being the first person on this untread path, seeing the land and space. It's intoxicating—you want to do more and more of it.

Looking into Duck Bay for the first time (figure 5.3) and then standing inside, before Cape St. Mary (figure 5.4), we can experience the same awe and mystery. What happened here? How long has this place looked like this?

Asked specifically about why the MER work is important, Cabrol echoes Rice's experience:

> This is exploration. I've never lost, even though we are remotely located now [not working together at JPL], this feeling of when we were doing the nominal mission. I was there three

or four hours prior to the download of the data or the beginning of the day, whenever the beginning of the day was. Which means that I would be there at 1 o'clock, 2 o'clock in the morning [Gusev time], usually alone, or maybe there was one other person, dealing with Pancam images, and the only other people who were there were the engineers. And there were two things that I remember. After a while, when we knew each other very well, the engineers running down to the science room and saying, "Hey, Nathalie, the data is coming down, do you want to come and see that?!" And I remember being there and just looking at something for the first time that nobody in the entire planet has seen before you—this is really exploration. And so, this is why this cannot become at any time routine, this is why I want to stay involved.

In imagining a revision to his book, Squyres thinks of it as "telling the rest of the adventure." The saga of MER is a journey of places, expected events, and emotions: "I'd describe the important discoveries that came after sol 250 or whatever it was *(laughs)* [when] I ended the book. And the adventures that we've had along the way: the dust storm, Purgatory Dune, the summit of Husband Hill, Home Plate, all of those, the descent into Victoria crater, just the feeling of getting to the rim of Victoria crater … all of that stuff, I'd want to get that adventure in."

In the process of exploring, the scientists' knowledge and methods blend further, as their objective is most broadly described as "figuring out the planet." Accordingly, Cabrol described planetary science as "the exploration of a system"; during the experience of developing and using robotic laboratories to study extreme environments, her geology interests blended with biology.

Sims explained how the Mars Exploration Rovers were designed to enable field scientists to approach this broader, systemic inquiry: "What we did differently on a planet for really the first time, much more strongly than Pathfinder, is that we were trying to really figure out the planet. We were doing what a field geologist or a field astrobiologist or a field geochemist would do. We were really trying to do exactly the same problems that they would have taken to the field."

Carr explained that this larger inquiry forces the scientists to relate to and learn from each other, broadening their identity:

But there's another aspect, of exploring a planet, something like Mars. If I were on Earth doing normal, conventional terrestrial geology, I'd have a narrow, specialized area that I developed to be an expert in. With Mars, you've got the whole planet! *(speaking highly animated, energized)* And you've not only got the whole planet, you've got all the engineering, the mission aspects. I think that's an added attraction in the sense that it's a collegial kind of thing. Most scientists work in their office, with their heads down, and their communication with other scientists is limited. But when you're on these missions, you've got to work with everybody else. You've got to put your heads together. You've all got to come to agreement and you get to know everybody. You get to know the people you're working with pretty well—not only the scientists, but the engineers, the whole thing…. I always found that aspect of astrogeology attractive. It's something that regular geology really doesn't have.

FIGURE 5.3
Opportunity arrives at Duck Bay at Victoria
crater on sol 952 (September 28, 2006).
Image credit: NASA/JPL-Caltech. *Source:*
http://photojournal.jpl.nasa.gov/catalog/
PIA08783 (accessed September 29, 2006).

FIGURE 5.4
Cape St. Mary of Victoria crater, 15 m promon-
tory exemplifying large-scale cross bedding
from ancient sand dunes. This Opportunity
"super-resolution" Pancam image was interpo-
lated from thirty-two images, sol 1213 (June 23,
2007). *Image credit:* NASA/JPL-Caltech/
Cornell University. *Source:* http://photojournal.
jpl.nasa.gov/catalog/PIA10211 (accessed
January 28, 2008).

Having technical people with whom to share the work—people with different interests and capabilities—enhances the experience. The science and engineering teams need each other, and they can share their excitement with each other and point out complementary ways to advance each other's goals, further motivating the work.

Culmination of a Career

The MER scientists talk about being on MER in reverential terms, as an avocation, something based in deep conviction and desire—almost like a mission in the religious sense.

The older scientists emphasize how MER brings together their experience, as if they have been on a path preparing them for this mission. Carr said, "I worked on shockwaves, chemistry, and cameras. Somehow I was prepped, and lucky … and I did all this fieldwork. If I had looked at what should an astrogeologist, a Martian geologist have, just by circumstance, I had a lot of it." But then he waited almost twenty years for this opportunity, a period during which he was "quite discouraged, of the possibility of ever seeing another Mars mission."

Similarly, Des Marais says, "MER represents a consummation of a career at NASA. It provides a degree of closure in many things I felt I was preparing for, on behalf of NASA, some twenty years before it happened."

For Squyres, the principal investigator, the MER mission represents a dramatic realization of a long-term effort. Including the initial conception of instruments in the late 1980s, persistent development of rover proposals in the 1990s, and being PI since project selection in July 2000, the MER missions will perhaps dominate Squyres's career. His interest began as an undergraduate when a friend helped him wheedle his way into a graduate seminar on Mars that provided access to the recently acquired Viking photographs. On seeing these images alone one day, Squyres says, "Sitting there cross-legged on the linoleum, I was exploring a new, distant and alien world." He realized that "geology on Mars, if there was a way to do it, would be something worth devoting a career to." A quarter of a century later, after a complicated and emotionally wrenching series of inventive proposals were rejected, delayed, and coerced into various configurations of landers and rovers, a team emerged that united previously competing instrument designers, and—through what felt like a miracle—Spirit and Opportunity were on Mars.[11]

MER is Cabrol's first mission and will forever mark her career. "The magnitude of what was happening was a very huge thing. For anybody to receive that phone call…." After working for so long to see the mission succeed, the landing of Spirit is a major personal accomplishment: "It was a very special moment, a very personal moment at so many levels. That is definitely a defining moment in somebody's career, somebody's life, seeing the first image." This feeling remains for years throughout the mission: "It is not something you can disconnect. Even when Spirit is done, it is something that will stay forever. It's something—and there will be other missions—but this one will stay the first one of this caliber, and it has been changing the face of planetary exploration. The team participating in that will share forever something very special."

MER is also Yingst's first mission, which she describes as "a dream come true": "You know, I'm just really a blessed individual, so fortunate. This has been a calling for me in the same way some people are called to be a preacher or something like that. And I am really, really blessed to be doing this."

Rice has participated in several Mars missions, but also starts by describing MER as part of a deeply held desire: "In terms of career satisfaction, yeah, it's pretty high for me, 'cause I'm doing something I dreamed about as a kid." He hopes to do more missions, but it doesn't matter; his life and his identity is now marked by MER, like winning a medal after years of struggling:

> That satisfaction is there and nobody can ever take it away from me. Everybody on the team's got these little NASA Group Achievement Awards, from NASA Headquarters. That's kind of neat. You can put it on your mantle one of these days. That's something, being part of this team, this experience I've had, memories. Yeah, it's very comforting and it's very fulfilling because that's something I scrounged around all my life, wanting to do, figuring out how I was going to do, didn't know if it was ever going to happen. I had a lot of disappointments along the way, with grad school and things, even in high school having some teachers saying I wasn't smart enough, I'd never work for NASA. It's about just not giving up, having people help you along the way, believing in yourself, to be part of this.

How Does My Expertise Fit?

Despite the unabashed excitement upon being selected, each of the MER scientists had to find a way to fit into the mission. Des Marais found it natural to "coordinate between disciplines" and became a frequent leader of long-term planning. Because it was a voyage of discovery, what the mission would require and how their skills would fit was partly unknown at first and developed over time. Sims, the engineer embedded in the MER science team, described this stage as simply trying to make himself useful. Rice relates what another team member told him: "Early on when Spirit landed, Grotzinger was thinking, 'What in the world am I going to do here? I don't know anything about volcanic rocks. Why did I get picked for this team? Because I'm not going to be of any use.' And then, lo and behold! Three weeks later, we land in Meridiani, and some of his experience and expertise is immediately valuable."

Sometimes the necessary expertise must be learned during the mission. For example, the MER instruments do not just send back pictures and charts. Everything must be processed and refined by the scientists, using a variety of computer tools. This processing requires a new expertise, as Yingst explains:

> But it's not the same thing when your instrument is not a hand lens—it's a very complex camera that gives you back information in ones and zeroes, instead of an image that you can immediately understand. A Pancam image, once it's downloaded—once it's sent back to us by the spacecraft—has to be looked over and calibrated. You have to know

what you're looking at. And then various versions of those images have to be released in formats that computer software can read and understand. That's a lot of background information to digest before you get to a picture of a rock, which is what you're trying to get to. And if you don't have a strong computer background, with doing spectroscopy or remote sensing in general, it can be very difficult.

The shift in instruments and expertise is particularly acute for Carr, who worked on every Mars mission before MER. He now experienced a disjuncture—working with a rover not only involves very different technology but also brings a dramatically new perspective: "I spent my whole career looking down from above. And now we're down on the ground. It's a very different experience, from what I had before. But, the thing of is, because all my experience was interpreting these big pictures, when we get down on the ground—this is me personally now—I'm a little uncomfortable at the narrow focus of the science." Being on the surface of Mars changes the nature of the observations and knowledge involved. Working with a rover shifts the study of Mars from broad investigation of the impact craters, volcanism, canyons, valley networks, and other large-scale features to detailed geomorphology and chemistry of classes of rocks (with specific names like "Clovis" and "Wishstone"). Compared to Yingst's experience, for Carr the difficulty wasn't just a shift in field tools, from a hand lens to the micro-imager, but a basic shift in the scientific topics, from tectonics and seasonal ice formation to detailed investigation of layered deposits, groundwater movement, and evaporitic sources[12]: "I'm simply not trained. For the really startling data, the chemistry, the sedimentary structures and so on, I'm useless…. Fortunately, we had on that team incredibly savvy guys, Grotzinger, Morris … they are really good guys. And I felt a little out of it. A lot out of it, actually." What about his geochemistry background? "Well, again, these are very good geochemists, people like Dave Des Marais. And there was a group from Long Island…. These are very good, smart guys. You know, my expertise is quite different; it's in the global relations of the planet. So I felt a little out of my field. And you know, I still do. You know, I'm 71 years old. Most of my life I've done things that are very different! (laughs)" Ironically, although Carr was in some ways unsure of his role, he readily recognized the relevance of the work to questions he been wondering about for decades: "Even though I felt a little out of place, the information that it's been giving us is so wonderful, and it helps me do the kind thing I've been doing all my life, the big picture stuff. It really helps knowing that there were indeed water-laid sediments there in Meridiani, and all the sulfur that's around, and so on. It changes everything; I can relook at everything."

Carr participated for the first nine months of the mission, often as SOWG chair (figure 5.5/plate 7), and then he retired. However, he's kept active, saying, "I check what's going on all the time." And he has reaffirmed his role in the Mars science community: "I've been writing another book and I've just about finished, and that's involved digging into a lot of the stuff. It's a replacement for my 1981 book, it'll have the same title, *The Surface of Mars*. Yeah, it's the whole thing, I mean I'm interested in the big picture, you know? But there's a chapter in there, quite a detailed chapter on MER, the science results from MER and how they fit into the rest of the system." So after all, Carr reaffirmed his niche in the community and his professional identity by assimilating the mission into the bigger picture, issues of interest to the entire planetary science community. Reflecting his global perspective and humor, the new chapter in his book (about Viking, Pathfinder, and the Mars Exploration Rovers) is ironically named "View from the Surface."

Yingst provides a counterpoint to Carr's recent retirement, as someone who joined the team as a participating scientist to replace others. She explained that the mission had been going on for so long, "some of the people who've been on the mission wanted to move on to do other things that they had already promised to do, and some people were just getting worn down." She was selected in November 2005 and then joined the telecons while continuing her academic and fieldwork (our interview was conducted by telephone when she was in the field, studying a gorge in Oregon).

Asked what it was like to join a team that had a history of several years, Yingst replied that this was indeed on her mind:

FIGURE 5.5/PLATE 7
Science assessment meeting (February 10, 2004). Mike Carr, SOWG chair, presides from the front center of the room (in white shirt near screens). Steve Squyres observes from the rear (image far left).

On the one hand, it's been great because people like Steve Squyres and the other MER scientists have been great colleagues. They've been very welcoming, and they have treated us just like we had been there since day one. And I have been very, very grateful for that. I've talked about this to the other new team members as well; we have really enjoyed it. It's been just an excellent experience. Having said that, of course, you know these folks have been doing this for ... golly, gee ... two, three years now! And so there's a camaraderie there, a shared history that we simply don't have. But we knew that going in. And while it's unfortunate that we can't have all of that shared past, that's okay *(spoken slowly, deliberately leveled)* ... as far as I'm concerned. Again, the important thing is that we're really have been considered full team members since the beginning and treated that way. And, as well, the new team members kind of have a little shared history of their own, which is nice, too.

I heard the term "newbies" when observing the science team in February 2004, but Yingst says new team members do not have a name for themselves: "We're all participating scientists, or we're all called Athena team members. There's no differentiation."

Asked what it was like to learn the new disciplines, the methods of handling the rover, the meeting protocols, and the tools, she said:

Oh, yeah, it was sort of "open mouth, insert fire hose." It was really intense. But if you went into this thinking, "I'm going to get all this and I'm going to be ready in a week to go out there and be a full participant," it's just not happening. But if you went in thinking, "I'm going to get as much as I can now. I'm going to jump in, and fully participate and do the best I can, and stumble along the way, and expect the people are going to be okay when I stumble," it was fine.

In summary, the MER scientists' training and their avocation as explorers led them to be part of the mission. All view the mission as a culmination of their career, with different realizations depending on where each stands in the personal timeline. But all bring to the mission a blended expertise that each must reinterpret and adapt in different ways into the actual work of "figuring out the planet." For Squyres as PI, the quality of the science could be tracked back to the design and testing of the six instruments and the software onboard the rovers. For Carr, the integration looks upward to the data observed from orbit and the broader concerns of the community reflected in a seminal text.

The next chapter considers in more detail the science operations and the reality of being on Mars within the body of a rover.

Working with a MER rover requires tending it day after day for years, planning and interpreting its actions as a team as it returns some images and spectral data, rolls to the next destination, or remains in a small area for months scrutinizing nearby rocks and soil. Individual contributions vary over time but the team remains engaged in the investigation. As Rice said in the summer of 2006, the experience of being on Mars is captivating, and new data continually invigorates their interest: "I'm still heavily involved in operations, and I'm geology lead for a week out of every month. Even when you're not the lead, I participate anyway because of the interest and honor of being here, to keep up with it and look at these pictures that are coming down every day. I mean, we're on the surface of Mars! We've got a continual presence on Mars—we've had it for over two years."[1]

The words of the MER scientists place them personally on Mars. In this chapter, I consider how the psychological perspective of "being the rover" facilitates using the mobile programmable laboratory to do planetary field science. My analysis relates how people talk, the different mental points of view they adopt, what it means for a person to be an agent on Mars, and the visualization and programming tools that make agency possible. My objective is to develop a more precise understanding of the MER exploration system—specifically, a crisp way of thinking about human action and robotic systems—so that we can better describe how telerobotic methods enable relating human intentions to programmed mechanisms, how the methods can be improved to facilitate human action, and what, if any, aspects of the work are left undone because they inherently require people to be physically present on the martian surface.

The Explorers' Language
People talk about the rover and the work on Mars in different ways for different reasons. We can characterize the references to what is happening on Mars grammatically as "first person" (I/we did something), "second person" (you, the rover, did something), and "third person" (it—Spirit or Opportunity—did something). These ways of talking reveal how people are conceiving (mentally coordinating) their relation to the rover and its actions.

When the scientists describe themselves as personally being on Mars, they are using the first-person perspective. For example, Rice said, "Meridiani was such a different landing site because *everything was so laid out in front of me*; there were obvious things to do." Wondering where the lakebeds of Gusev might be, "People started saying, *it's below our feet*." Asked how he visualized his work, he said: "I put myself out there in the scene, the rover, *with two boots on the ground*, trying to figure out where to go and what to do, how to make that what we're observing with the instruments. By and large, day in day out, it was always the perspective of being on the surface and trying to draw in your own field experience in places that might be similar—how you'd detect a landscape and interrogate it as much as you could." Squyres's book *Roving Mars* is full of references to the team being on Mars: "As we work our way across the plains"; "We've arrived at Endurance crater"; "The slope immediately in front of us"; "Where we're standing now."[2]

The scientists specifically talk about *moving* themselves on Mars. For example, even before beginning the study of the outcrop in Eagle crater where Opportunity had rolled to a stop on landing, they felt the urgency of getting onto the plain: "There were questions that could only be answered by climbing out of Eagle crater and exploring."[3]

Landing in Gusev, the team looked around, found their bearings, and appraised their opportunities. Squyres describes their perspective and eventual joy of climbing the hills: "I remember we landed at the Gusev site and we looked at the Columbia Hills and we joked about it, we laughed. 'You know, wouldn't it be cool if…. I wish we'd landed there, so we'd have a chance of scraping a little bit up that hill.' Yet on sol 600 we were on the summit!"[4] These projections of the self as being on the surface of Mars are enabled and often occur through the perspective of images, as Squyres relates about the landing of Opportunity (italic emphasizes the first-person orientation of being on Mars):

> I remember when *we landed with Opportunity*, the very night that *we landed*, we figured out more or less *where we were*. The guys at MIPL[5] found the MOC [Mars Orbiter Camera] strip that had the landing site in it and printed it out. It was like 20 feet long, and 3 feet wide, and I remember we rolled it out on the table *(sound effect)*. We realized *we had landed in a crater*, probably Eagle crater, and that's where we were. And then *we noticed, 800 meters away*—which is more than our 600 meters, but maybe we can make it—there's Endurance crater. We're talking about that, "My god, wouldn't it be great to actually get there!" And then, way down the table *(laughs)*, you know, the other end of the roll, there's this monstrous, crazy crater. We just laughed. "Oh, geez, wouldn't it have been cool if we'd landed down there?! Oh, well." *(Slaps hand on leg)* We've *been there* for, what? A year and a half now? At Victoria crater. So, at the time, none of it seemed attainable. But if you had told me that *we were going to climb* to the summit of Husband Hill, and get to Home Plate, then get down into Endurance crater, out and to Victoria crater … I just wouldn't have believed it.

Strikingly, many of these first-person references are to the group of scientists as a whole, as if they are standing, moving, climbing, and looking together. Yingst mentions that each day the group must decide, "Well, *are we going to go* or *are we going to stay here* and take more pictures or take more data?" Vertesi has pointed out that saying "we" focused the multidisciplinary team in the nominal mission and later became a resource for bonding the distributed team.[6]

Much less obvious to people who are not living through the rover's actions is how the scientists experience changing the surface of Mars from a first-person perspective, that is, as actors. The scientists can see the RAT circles they have created and how they have disturbed the soil (figure 6.1). Looking at the resulting images, the scientists can view the marks as if their own boots scuffed the ground. Together with the engineers, they envision certain movements that are translated into specific rover commands.

There is another first-person way of seeing an image like figure 6.1 that anyone can experience: put yourself back on Earth, and consider, right now, as you sit here and contemplate this photograph, that these tracks and scuff marks are out there, on Mars.

FIGURE 6.1
Tracks deliberately made using Spirit as if scuffing one's feet, described as choreographing "an intricate dance ... maneuvering it up the side of the dune, shimmying its left front wheel a number of times to create the scuff, and then reversing to attain proper positioning" for Mini-TES observations, on sol 72 (March 17, 2004). *Image credit:* NASA/JPL. *Source:* http://photojournal.jpl.nasa.gov/catalog/PIA05577 (accessed June 26, 2009).

We have been there and we did this. These are our marks—our boots on the ground of another planet.

The second-person perspective, the rover as a partner, is far less common. In the interviews, it was only expressed by Trebi-Ollennu, a lead rover planner. His perspective of working with the rover in a joint activity is detailed subsequently in the section "Coordinating Telerobotic Inquiry: Visualization and Autonomy" later in this chapter.

The third-person perspective, the rover as an external, independent agent that we are observing and talking about (figure 6.2/plate 8), is the conventional shorthand that has appeared in print since at least the days of Viking. For example, in 1984 Ezell and Ezell wrote, "Viking's explorations and discoveries did not stop with the search for life."[7] As explained in chapter 1, the MER scientists use this convention in their formal publications. *The Planetary Report* might be viewed as informal writing, but even the team's *Science* 2009 article begins, "The Mars Exploration Rover Opportunity examined a small bedrock outcrop at its landing site in Eagle crater." The fieldwork is described from the rover's perspective throughout—Opportunity "entered," "drove," "encountered." And the scientists summarize their work by saying, "Opportunity's investigation of Victoria crater shows...."[8] The section "'Robotic Geologists'" later in this chapter examines the value and confusions in this elision, which veils the scientists' agency.

Figure 6.2 illustrates how we exploit and invent computer graphics to create synthetic third-person views, which were not so easily (or convincingly) generated in the time of the Vikings. The image of Opportunity on Burns Cliff is an example of Virtual Presence in Space technology, which combines "visualization and image-processing tools with Hollywood-style special effects." This "you are there" perspective, with the rover denoting the scale, places the viewer on the surface of Mars, corresponding to the scientists' mental image of a landscape laid out before them. In stepping back to "see the rover" in this way, MER also becomes objectified as a third party—an independent actor on Mars.

In summary, different perspectives—we are on Mars, the rover is working with me, and the rover is on Mars—are useful for talking about the MER mission, either in doing the work or reporting what has happened. The perspectives get blurred however, even in scientific articles, and this can confuse our understanding of what we can accomplish with telerobotics and what radical technological improvements are actually required to develop a human-capable, robotic explorer. To this end, the following sections examine the notion of first-person embodiment in the rover, visualization methods, agency, and the third-person metaphor in more detail.

The Rover as a Human Body

The scientists speak about the rovers being their surrogates because its sensors and effectors were designed to enable them to perceive and act in ways important to doing field science on Mars. In practice, the robotic laboratory serves as the virtual body of the entire science team. The team uploads its program of behaviors each sol: the "sequence,"

FIGURE 6.2/PLATE 8
Simulated image of Opportunity superimposed on Burns Cliff in the inner
wall of Endurance crater. The view combines forty-six frames taken
by Opportunity's panoramic camera between the rover's 287th and 294th
sols (November 13–20, 2004). *Image credit:* NASA/JPL-Solar System
Visualization Team. *Source:* http://photojournal.jpl.nasa.gov/catalog/
PIA03241 (accessed January 25, 2008).

which details what instruments will be used where and for how long, as well as when the rover will drive or be idle (to recharge the batteries from solar power, referred to as a "siesta"). Thus, the aspirations and personal inquiries of dozens of scientists are each sol interleaved and reduced to an eight-hour linear program (figure 6.3).

What is it like to do field science as a team using a rover in this manner? Rice describes a bittersweet experience:

> Frustration comes to mind; they're so slow and plodding. Think what I can do in the field versus that rover.... But that's unfair.... These things have been our eyeballs out there and our legs and our arms. Working with these rovers? It's been some kind of weird, man-machine bond. *(laughs)* It's become an extension of each one of us, our eyes or our hands, our feet. And I guess in a way, tasting the rock. It's kind of like it's morphed into us, or we've morphed into it. But at the same time there is that frustration. With Spirit, we can't go anywhere right now because it's the winter season. Even when we're driving, these things are going pretty slow. And you just want to just hop over those rocks or hop over that ridge over there and climb it, bang on it, do things.

Asked to describe her relation to the rover, Cabrol said, "What about symbiosis?" So are you part of the machine looking out and reaching? "Well, you have to imagine yourself in the field—what would you do if you were there? You have to be totally in tune with your rover in knowing its capabilities so that you know what is the next critical observation that you have to do."

When imagining the rover's position, such as with its IDD perched on the inner wall of Victoria crater (figure 6.4), does Des Marais see the rover as "this thing that's out there, in this place" (third person) or, in visualizing the surface, does he see things from the point of view of being the rover (first person)?

> I was giving talks about this in the first year or so, and I would talk about what the rover was doing: "This is the rover's arm" *(gestures)*—I was standing up there on the stage almost acting like the rover—"and then we turned." I think you have to project yourself into the rover a little bit: "There's something over there of interest to us ... 'something I thought about yesterday.' Can we still see it? What can you see if you looked over there? How far out *(gesture)* does that solar panel on the rover sit, and does it block your view?"

It as if you are the rover itself, looming over the rock surface, with wheels, a metal body, and an IDD (figure 6.4). On imaging the El Capitan outcrop, Squyres described the same embodied "you are there" perspective: "What would you do if you were there? You'd get down on your hands and knees, right?"[9] So rather than look with the Pancam, elevated off the ground, they used the MI on the IDD.

Des Marais explained that this visualization process includes not just what's reachable or visible but also the lighting, which changes during the day on Mars. With time, these simulated projections in the imagination became more automatic:

09:15 seq resume time
09:30-10:10 HGA 20240 @1000/16590 (15min warmup, No cal)
10:05 master handover
10:10 morning remote science
10:32-11:30 Snooze
11:30-12:30 IDD stuff (MI, MI, MI, APXS door open, MB place)
12:30-14:00 Remote Science
13:30-14:00 PM HGA 20241 @1000/16590
14:05-16:39 Siesta
16:47 ODY 40241 – MTES during
01:43 MGS 30250 – Shutdown @ 02:03:30.
02:26 Wakeup –Collect MB data; (02:31) turn on IDD heater.   Sleep again
03:00 Wakeup – Turn off IDD heat; do tool change, start APXS. (must be done by 3:14).
03:22 ODY 40250 – (03:14 to 03:37 full window) Do RPDU switch eng. sequence (e0102).
03:42 Shutdown

FIGURE 6.3
Timeline for sol 24 of Opportunity, also called the "command sequence,"
as reviewed at the science context meeting on the morning of sol 24
(Mars local standard time). Work is constrained by power and communication appearing here as operations involving the High-Gain Antenna (HGA)
and the satellites then in Mars orbit serving as communication relays,
Odyssey (ODY) and Mars Global Surveyor (MGS). Instruments to be used
on sol 24 are the Microscopic Imager (MI), APXS, Mössbauer Spectrometer
(MB), and Mini-TES (MTES), which are deployed by the robotic arm (IDD).
The planned four rests create power reserves in anticipation of a drive on
sol 25 (later deferred to sol 26).

Also, in a number of observations, your angle with respect to the sun is really important for what you're going to try to do. Is this rock going to be in shade or is it in the sun? Positioning yourself with respect to the sun outdoors, you don't even think about it—it's intuitive. For Mars, you want to get to a point where it's intuitive. We didn't have a lot of time. We had to do all these decisions, and I think it forced us to project ourselves into the rover. The first few months of the mission, they had these huge charts on the wall, engineering drawings of the rover, with all of the dimensions. We'd have some geometric question, "Well, can we see this; can we reach this?" We'd go stand and stare at those charts, and over time we stopped doing it so much because we began to gain a sense of the body. That's definitely projecting yourself into the rover. It's just an amazing capability of the human mind—that you can sort of retool yourself.

These projections—"inhabiting the rover"—occur especially as the scientists formalize and visualize plans for the next day in computer emulations of the rover. The Science Activity Planner (SAP) program (an improved version was called Maestro) enables commanding the rover by pointing to and labeling images previously taken, such that chosen targets are automatically registered in precise three-dimensional Mars terrain coordinates (figure 6.5/plate 9).[10] Other programs, such as the Rover Sequencing and Visualization Program (RSVP), then convert these instrument and target sequences into specific movements and orientations of the arm and instruments.[11] Through this tight coupling of image, targeting, and feedback, plans are transparently enacted into exploration paths such that the scientists move over several days from broad panoramas to outcrops to particular rocks and then a handful of sharply resolved grains. As the rover moves forward, the returned images are ever more detailed, and the scientists zero in on their interests and understanding of how the rocks and terrain formed over the millennia.

FIGURE 6.4
Opportunity Hazcam image on sol 1,305 (September 25, 2007) at bright band of rocks in Victoria crater. *Image credit:* NASA/JPL-Caltech. *Source:* http://photojournal.jpl.nasa.gov/catalog/PIA10006 (accessed January 28, 2008).

FIGURE 6.5/PLATE 9
Image overlaid in the Science Activity Planner with target names and bounding box for candidate images (February 19, 2004).

Indeed, the use of simulation is so prevalent in the MER mission that it appears, as Des Marais says, even in taken-for-granted informal ways—in their gestures and imaginations. Working also with a duplicate rover in a simulated Mars terrain at JPL managed by the engineers, the scientists further simulate how the rover will behave—compensating for the impossibility of directly touching, seeing, or manipulating the stuff of Mars itself. In coordinating the rover's work across a variety of physical and computational models, the projection of the self into the rover is an embodied way of synthesizing these disparate sources of information: as Cabrol says, "What would you do if you were there?"

Schön emphasized how professionals—particularly, designers and engineers—effectively coordinate developing concepts with physical artifacts and models through a reflective and manipulatively iterative process of "seeing as" and "conversation with materials."[12] That is, reasoning involves a tight feedback between interpreting and trying ideas. In MER, this occurs both through a combination of imagination, computer-based visualization, and physical mockups.

Des Marais refers to retooling himself, Rice to "morphing," and Cabrol to a symbiosis—the team has become the rover. Apparently, the capability to move accentuates this feeling. Simply being a "lander" on the surface, such as Viking in 1976, gives some sense of presence. A rover is different, says Rice, "because you can move. When you go somewhere on vacation, you don't sit in one spot; you walk around, go to a museum or whatever. The rover aspect really drives home [its value for fieldwork]; it's the next best thing to being there." The almost full-size panorama photographs laid out on the table of the science assessment room provide a way for scientists to walk around seeing and pointing as if they were the rover on Mars and planning where they would like to go and which areas they would like to scrape and analyze (figure 6.6/plate 10).

FIGURE 6.6/PLATE 10
Examining panorama of Eagle crater outcrop in Meridiani science assessment room (February 18, 2004); Rice is leaning with his hands on the table.

CHAPTER 6

Working from orbit is different yet again; as Rice says, "You're more removed and remote." It is this shift in perspective—a different way of visualizing Mars, of projecting himself onto the planet through the rover—that Carr, with his extensive experience orbiting Mars, found problematic. With the shift in view comes a shift in what morphological clues and processes the scientists are studying—such as the shift from apparently eroded river channels viewed from orbit to a patch of silica-rich sand stirred up by chance in the rover's tracks. A scientist's way of seeing and way of thinking are coupled together in experience; different ways of seeing can't be trivially or immediately mapped onto each other.

Nevertheless, being in orbit provides its own "you are there" experience. Describing the Cassini team's experience cruising around Saturn, Meltzer says, "We're all standing on the bridge of the starship *Enterprise*."[13] As a student participant on Voyager 2, Jim Bell recalls "watching the images come down, one by one, as each planet got bigger and bigger in the windshield."[14] The effect of "being there" has been experienced by mission scientists throughout the history of space exploration, as described by Oran W. Nicks in 1985:

> While I have often thought of these missions as similar to the expeditions of great explorers like Lewis and Clark, one beautiful consequence of today's communication systems is the immediacy of sharing the experiences, findings, and results of exploratory missions. I have heard a number of scientists express the feeling of "being there" with Mariners, Surveyors, and the like, for they could associate with those lifelike machines, superposing their own human characteristics, without having to consider "the other human being." Thus, our automated explorers have truly been extensions of man in fulfilling our exploratory desires in a briefer span of time and with broader participation than otherwise would have been possible. They have allowed us experiences that in the past would only have been available to the hardy explorer, without our risking life, limb, and personal resources.[15]

Nicks mentions all three relational metaphors: the spacecraft as a means for first-person embodiment ("extensions of man"), a second-person empathetic association ("partners to man"[16]), and a third-person automated explorer (as in his book's title, *Far Travelers: Exploring Machines*). As Nicks says, perhaps the Mariner and Surveyor scientists were there *with* the machines, but the MER scientists are there *as* the rover, embodied within it. Or maybe by "superposing human characteristics," Nicks meant first-person embodiment, too, in which "the other human being"—meaning the lander or orbiter—becomes peripheral. Regardless, a telerobot is more than a mere "extension of man," in the manner of a pole, a jackhammer, or even a backhoe. A telerobot like MER is a means for presence, for acting on the surface, by enacting a customized procedure (the daily command sequence).

To get a sense of the embodied experience, imagine for a moment being an elephant, with your heavy legs and proboscis swinging and curling like an arm. Now imagine that you are controlling this elephant body to move and manipulate things remotely (see figure 6.1). This is what we mean by "embodiment": you take on the characteristics of the machine, like you're stepping into a suit and acting from inside it.

The experience of being effective at a distance through a telerobot is called *mediated agency.*[17] People who are not members of the MER science or engineering teams might not feel either the sense of agency (acting on Mars) or presence (being on Mars) because they have not experienced "becoming the rover." They have not been routinely interacting with the landscape over a period of years: reaching out the rover's arm, inspecting and manipulating materials, moving through sand, holding tenaciously to steep crater walls, and refining their vision through increasingly detailed photographs and micro-images.

Although the scientists' proximal physical interactions are often performed via computer keyboards and pointing devices, they are *conceptually interacting* with rock and soil surfaces within a visible landscape on Mars (figure 7.2). As agents on Mars, the scientists' actions occur in a conceptual, visually imagined space—not restricted to their immediate physical surroundings on Earth. As they say, they are conceptually situated, that is, imagining places, movements, and changes of a visualized martian environment. To sustain the exploration as the environment and state of the rover change (e.g., swirling dust obscuring the light, a wheel breaking, a temperature sensor stuck), what is happening on Mars must be reimagined, models revised, plans adapted, and new ways of using the available capabilities invented.

Agency requires not merely the ability to sense and act (as the rovers do through the uplinked programs), but to *perceive* (categorize sensations and postures, i.e., physical body positions) and *conceive* (categorize relations between perceptions and actions, e.g., "scuffing"). Thus the experience of agency (being an effective actor) and embodied presence (being the body of the rover on Mars) go together, as occurs in a person's experience of video games and virtual worlds on the Internet: "Presence depends on a suitable integration of aspects relevant to an agent's movement and perception, to her actions, and to her conception of the overall situation in which she finds herself, as well as on how these aspects mesh with the possibilities for action afforded in the interaction with the virtual environment."[18]

Of course, for MER the environment is not virtual (an artificial reality), but actual (a place on Mars). Unlike telemedicine, in which a surgeon might act in a different place at the same time, the scientists act at a different place at a different time (figure 6.9), a distinction made possible by representing their actions in computer programs that will enact their intentions during the next sol. As explained in chapter 4, daily commanding facilitates the sense of being there, of being an agent on Mars. In this respect, one can view the combination of the robotic system and programs as constituting a "surrogate"—not substituting for the person, but *acting as the person directed.* Simply put, MER is a *physical surrogate*, not a cognitive surrogate. The rover planners can refer to the scientists' annotated "intentions" in the observation/activity plan, but MER cannot interpret plans in this manner; MER only executes the plan (which can specify options and criteria for gathering and selecting data depending on resources and circumstances).

As a physical surrogate, the MER telerobotic system is also analogous to a messenger, who simply relays what was already said, as opposed to an ambassador, who as a representative of a government must articulate policies and investigate new ways to realize the goals of his country (a *social* surrogate). On the other hand, like a messenger, a telerobot is much more than a puppet: it must be able to move to designated locations, finding new routes if obstacles are encountered, and to preserve itself when undesirable conditions occur—though these capabilities are today extremely limited, as incidents such as getting mired in sand attest. Lunokhod was more literally like a puppet, joysticked from Earth; Surveyor, a fixed lander, was similarly operated by remote control, that is, sending commands to immediately operate camera, arm, and instruments.

In summary, the MER scientists have the experience of being effective agents on Mars because their perceptions of being present, their conceptions of their inquiry (models, methods, plans), and their resulting actions are coordinated. Everything fits and works together, conceptually coordinated, in their experience. The ability—through cognitive, social, and technological processes—to decide about actions that will occur as they intend is crucial to the experience of having agency on Mars. The rovers lack this ability and therefore cannot carry out a scientific inquiry on Mars (compare figure 3.6).

The idea of a *cyborg* provides another way to understand the fine distinctions possible between an "extension" that enhances human agency (Nick's term) and a "surrogate" that *enables* human agency. A cyborg can be defined as a human with machine parts, usually implanted in the person's body,[19] such as prosthetic devices (e.g., insulin pump). Some extensions of the body incorporate real-time internal feedback that integrates their actions with the body's processes. Depending on the degree of contribution, such devices extend (augment) or replace body functions. By this definition, a person with eyeglasses is a cyborg.

Technically, the MER scientists are not cyborgs by virtue of using MER. The rover does not functionally augment the body in real time (an integrated extension of the body) but is rather a device that enables people to have agency on Mars. However, some might view MER as a bodily extension because the images it provides, combined with visualization and programming tools, provide a conceptual illusion that action is not mediated,[20] that we are ourselves acting on Mars (figures 7.2, 8.5, and 8.7). Furthermore, people are adapting to the rover's sensors and effectors; they think in terms of what they can do on Mars with these ways of looking and manipulating. Insofar as using these interfaces becomes automatic to the scientists and the rover's capabilities are "unnatural" (beyond what we can normally sense and manipulate), the scientists might find themselves adapting to the machine's capabilities in the kinds of sensing and manipulations they can imagine—what Des Marais called "retooling." Once you begin to think in terms of seeing with Mini-TES eyes, you are effectively working as a cyborg.[21]

During Spirit's sol 37, I walked over to Rice at his workstation in the LTP area. In contrast with the just-completed science assessment meeting, this was an incredibly fun forty-five minutes. Rice was reviewing recent downloads from Meridiani that I hadn't seen either (ironically, we were sitting in the Gusev team's room). Looking at images on the screen was like being at that spot on Mars together, looking at details of the fine lamination and weathering. Rice pointed out several times how the process was caught in different phases, with the eroded friable material wearing down to a more solid unit resembling pavers. The best part was when we viewed stereo images wearing red-green glasses and pointed at various things, describing them and what they were. Rice was having trouble getting part of the image to fuse into a three-dimensional perspective, so we sat back, and it was like looking through a window together and really being in the scene (see figure 6.7/plate 11). There was no real difference from being on Mars and seeing this together for the first time (except, of course, in terms of safety and convenience). With our 3D glasses in place, we were fusing our bodies with what the cameras imaged on Mars, so maybe we were becoming cyborgs after all.

FIGURE 6.7/PLATE 11
Scientists and JPL engineers examining a 3D image in the science assessment room (February 19, 2004). Standing from left to right: Chris Leger, Bob Anderson, Andy Knoll, Tim Parker, and Jim Rice; Mike Malin is at the desk, driving the image.

CHAPTER 6

Coordinating Telerobotic Inquiry: Visualization and Autonomy

We have seen that the MER scientists' experience of presence and of being agents on Mars is facilitated by a combination of *mediated perception* (i.e., visualization tools, including MER's sensing instruments, analytic software, displays, stereo glasses) and *mediated action* (i.e., programs that control robotic tools such as the IDD, instruments, and the vehicle).[22] Perceiving and acting depend on each other and must be coordinated both on Mars and in the scientists' and engineers' activities on Earth: (1) *sensing* the Mars surface by moving and adjusting the instruments; (2) *interpreting* and *deliberating* by manipulating computerized renditions of data, such as photographs, Mini-TES overlays (figure 6.10), and graphs (figures 1.1 and 5.2), and (3) *commanding* subsequent observations (creating an upload sequence) by representing interpretations and intentions within visualizations (figures 6.5, 8.5, and 8.7). These three activities form a daily cycle of telerobotic activity, evaluation, and programming that becomes the coordinated activity of the entire team, constituting the rhythm of the work. In this section, I discuss the visualization and control issues in more detail; the next chapter explores the communal aspects.

Yingst gave some striking examples of how the scientists visualize the Mars surface, revealing how they orient themselves from the rover's standpoint:

> When you get a chance, look back at some of the pictures from Viking. You will see the most circumspect, quiet guys, with their hands folded across their chest, in the background, standing on their toes! Because they really want the lander to just, kind of look over, and they're doing it unconsciously. But, yes, absolutely, the rover's our proxy. And most of us would be there if we could. We do get that sense of being there. Sometimes you'll see people talking about "getting a picture behind us," and you'll see them turn their heads. Again, this is totally unconsciously, because they're thinking of themselves, if they were in the field, what they would try to be doing, what they would be attempting to look at.

As Des Marais mentions, knowing what the rover's cameras and instruments can "see" and where its instruments can reach pose extremely complex and essential problems for daily programming. Generally, photos from previous days are used to locate the rover precisely. Yet these pose other problems in interpretation and 3D visualization. Referring to Opportunity's hole-in-one landing (figures 3.4 and 6.8), Rice said: "It took us a while to figure out that we were in a crater. Early on, we were calling that outcrop the "great wall." We didn't know if we could drive over it. Turns out, it was only like a little curb, 10 or 20 centimeters high."

Just as if one were on Mars, it was necessary to look again, shift perspective, and move closer to see what was really there. Whether moving an arm (MER's IDD and the shovel arm on Viking and Phoenix) or the entire vehicle, a balance must be found between efficient action and not harming the machinery. Further, robotic actions are not always predictable in detail (such as how the rover will navigate a path or where the IDD will make contact with a surface), but the overall rules of behavior and targets can be constrained.

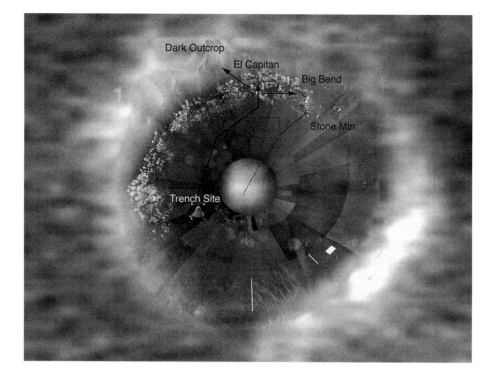

FIGURE 6.8
Manually composed rendition of Eagle crater landmarks and rover path.
Combines an MGS Mars Orbital Camera image and bird's-eye view of
Opportunity images generated in SAP. Dashed lines indicate the rover's
path to the Trench Site from the center landing spot and subsequently
Stone Mountain, with planned movement (solid line) to El Capitan and then
beyond to either Big Bend or the Dark Outcrop. *Image credit:* JPL/NASA/
Cornell/MSSS/Ohio State University. *Source:* JPL MER Mission, "Press
Release Images: Opportunity," http://marsrovers.jpl.nasa.gov/gallery/
press/opportunity/20040217a.html (accessed February 25, 2005).

Learning to program and use a telerobot in this way has been researched in artificial intelligence laboratories since the 1960s, but MER represents the first opportunity for long-range, long-duration programmed robotic investigations of a planetary surface.

Sims explains that in figuring out how to operate rovers so scientists can understand Mars, they are "inventing planetary science": "How do we learn with our machines, the synergistic combination of what they're good at and what we're good at?" Although the years of operating MER may appear easy to the public, how people should work with a robot has not been obvious to academic computer scientists or planetary scientists and varies considerably. Earlier, we compared "batch" commanding on the Viking and MER missions. But the difficult conceptual progress occurred in moving from direct, remote, and fixed commanding methods to the batch method (table 2.1) and then knowing how to practically control what will occur at another time and place to get the desired scientific results, without have to specify every action in advance.

In particular, according to Sims, researchers were at first fixated on controlling a robot by joysticking, the teleoperation method used by the Soviets to control the Lunokhod rover on the moon in the early 1970s. Consequently, engineers misconstrued the main problem in remote operations to be the time delay.

Operating Lunokhod involved a six-second round-trip delay between the transmission of an image from the rover and the receipt of an Earth-based action command by the rover.[23] Sims concluded from analog experiments that the real issue for Mars operations is not perceptual-motor coordination, but keeping track of what you are doing in terms of managing the people and the work process:

> When does latency get so bad you lose context? You lose controllability of the vehicle.
> How do we organize our science teams, our controlling teams, so they can manage
> a forty-minute time delay? So you can drive every forty minutes. Every forty minutes,
> we get a new command back, and we send up another one. And you do that all day.
> That's the Lunokhod model, which virtually everyone had about driving robots on Mars,
> prior to Pathfinder.

Using this brute-force "joystick" method of control with a forty-minute delay leads you to forget details of where you are and what you were doing: "We were in Kilauea and [the astronaut] Pete Conrad flipped over a Marsokhod,[24] driving it backward because of not being able to see it. We'd have an hour and a half discussion about what was around the rover, because you had no memory, you couldn't see it, it wasn't there. You were playing in Mars mode, so you had forty-minute delays. You couldn't get data back immediately; it wasn't like you could teleoperate it."

Joysticking a rover doesn't work with such time delays. To break out of the "direct manipulation" perspective, using rovers like hammers (direct extensions of the body), we had to understand that the real difficulty in working with rovers was personally placing ourselves in the body of the rover, in which it becomes our physical surrogate, and being able to quickly and precisely reorient ourselves when returning to the work hours or days later. Rather than controlling a machine "out there," you had to get inside the rover and become its mind in planning an entire day's work. Virtual reality tools provided the means to do this, as Sims explained:

> It turns out serendipitously that what we invented to control robots in artificial intelligence research turned out to be useful. We created an environment—virtual reality control of robotic devices. Otherwise, you didn't know what you were doing. You repeatedly forgot what was behind you. You couldn't keep track of what was on your right side. You could look at it, but where was it? What was over there and where are things? So we invented this idea of the virtual reality interface. My evaluation was that the most critical place for autonomy was virtual reality, much more important than planning programs, much more important than scheduling programs—much more important than everything else that we were doing in robotics.

What does Sims mean by "autonomy"? He provides a human-centered, comparative definition: "System A is more autonomous than system B, if in A humans do more of what they want and less of what they don't want. By that definition, virtual reality adds to the autonomy of the system." Using virtual reality tools to aid in planning the rover's actions allows the engineers to program the instruments at a higher level of abstraction (e.g., by pointing on a screen), leaving the precise movement calculations to software. Similarly, MER has autonomy in long drives by using photographs to detect obstacles and plan drives around them. In sufficiently smooth terrain, the scientists can just specify a distance and bearing and leave the details of navigation to the rover. These capabilities have been incrementally improved during the mission:

> In April 2004, the rover's software was upgraded to allow the rover's suspension angles to be checked against preset limits … thus enabling the rover to stop at negative terrain features (i.e., holes) that were not visible a priori. Because the reason for halting a drive (e.g., timeout, suspension check, slip amount, or tilt check) is accessible to the rover sequencing language, a recovery maneuver could be performed whenever the suspension check tripped. The recovery consists of backing up several meters and continuing the drive with AutoNav [Autonomous Navigation], since AutoNav is able to detect and avoid negative hazards.[25]

For example, after first uploading this software to Opportunity, the rover drove "55 meters (180 feet) … as a 'blind' guided drive [programmed by a rover planner] based on images acquired previously. Speed during that session averaged 120 m (394 ft) per hour." For the next 85.9 m, the rover switched to "autonomous navigation, watching for obstacles, choosing its own path, and averaging 40 m (131 ft) per hour."[26] With the AutoNav program, MER stops every ten seconds and then assesses the terrain for twenty seconds. Thus there is a trade-off between rover time and engineering time—the rover drives three times faster when the rover planners program the route in advance. Driving "blind" on a programmed route, the expected top speed on a flat, hard surface is about 3 m/min, and the achieved maximum was actually 183 m (600 ft) per hour.[27]

Other MER software "autonomy" upgrades include calculating how far to reach the rover's arm out to touch a rock and examining sets of sky images to transmit only those that show clouds or dust devils.[28] As a further step in automating decisions for acquiring data, the AEGIS (Autonomous Exploration for Gathering Increased Science) upgrade to Opportunity's computer in 2010 enables it to choose rock targets: AEGIS "examines images that the rover takes with its wide-angle navigation camera after a drive, and recognizes rocks that meet specified criteria, such as rounded shape or light color. It can then center its narrower-angle panoramic camera on the chosen target and take multiple images through color filters." Target criteria are programmable by the scientists: "In some environments, rocks that are dark and angular could be higher-priority targets than rocks that are light and rounded."[29] Marking the significance of this achievement, AEGIS received NASA's Software of the Year Award in 2011. In general, capabilities for automated navigation, planning, and data acquisition that have been demonstrated in robotics research go well beyond what can be practically exploited on MER, primarily due to the limited computer resources.[30] (MER has a 20-MHz RAD6000 CPU with 128 MB of RAM and 256 MB of flash memory;[31] using AEGIS requires up to two minutes to recognize a target and process another image.)

In summary, rather than being an inherent property of technology, such as software or a robot, autonomy is a relation between people, technology, and a task-environment (e.g., the terrain). Sims's definition accounts for human preferences—"more of what they want"—and is comparative—"system A is more autonomous than system B." Because people are inevitably involved in directing and monitoring even the most automated systems, this perspective makes clear that automation is for human purposes. Just making a vehicle more "independent" doesn't necessarily satisfy the needs of the scientists and engineers. Autonomy is valuable when it enables people to "do more of what they want to do." Put another way, using some forms of automation, scientists and engineers could lose their own autonomy—their ability to act intelligently with respect to what they value. Until rovers can set scientific goals and adapt methods intelligently, providing flexibility under human control to enable people to work more efficiently is paramount.

Trebi-Ollennu, the lead engineer for the MER instrument arm and a rover planner, also has a relational perspective about autonomy. As a planner, he interacts with the scientists, considering what they want to do, and then plans (programs) the drive and robotic arm activities. In a sense, the relation of scientists to the rover planner is like the relation of the rover planner to the rover: by offloading the targeting and traverse details to the rover planner, the scientists can devote more time to analysis and strategizing. And similarly, the preferred moniker "rover planner" (as opposed to "rover driver") reveals how engineering software has enabled offloading control of the IDD and navigation to the rover so that uploaded commands can be more abstract.

Given his role as intermediary in control of the rover, does Trebi-Ollennu visualize the rover as a third body, an object over there that he's manipulating, or does he project himself into the rover, as the scientists do, to imagine that he is there on the surface of Mars, looking and reaching? After all that the scientists said, his response is at first surprising:

> That's a very fascinating question because I look at it as a team. I actually see the rover as an equal partner. And I look at it as an "adjustable autonomy." The key thing here is that I have my limitations as a rover planner—resources, my time to build a sequence, then time to do the analysis. If I had all the resources in the world, I could do everything myself. I can build every little thing that I want the rover to do—intricate sequencing—so that the rover doesn't have to think for itself. But because of that limitation, I'm forced to work in partnership with the robot.

In effect, Trebi-Ollennu is restating Sims's principle. He adjusts the rover's independence according to what he wants—and is able—to do:

> Let's take a drive, it's always a very good example to use. If I want to drive 30 meters, I might have imaging to do surface mobility analysis for only 15 meters. So it's efficient: I can say to the rover, "Drive blind. Don't use any of your autonomous systems on board, just go straight there." It just turns on the wheels, it goes there—driving the rover blind is more efficient in power and time. After 15 meters, because I cannot see any more, I can say to the rover, "After 15 meters, turn on your autonomy, and you're on your own." Now it's taking images, taking about three minutes a step to analyze the image and do everything. In this example, we're working as a partnership. My limitations are being covered by the rover's abilities to do certain things that I cannot control.

From Trebi-Ollennu's perspective, each task can potentially be accomplished by a combination of his actions and the rover's actions. The optimal combination uses the information he has available and minimizes his planning time while also minimizing both the power and time required by the robot. When Trebi-Ollennu can see far enough ahead in the images, experiencing sufficient situation awareness of being on Mars, he adopts the first-person "I am the rover" perspective. Befitting the rover's dependence on the person, it is instructed to drive blind: when "I know everything, I just say, 'Go there.'" Significant examples of these combined programmed-blind and autonomous drives occurred on Opportunity during 2005:

> On sol 360, Opportunity traversed a record 154.65 meters (507.4 feet), using a combination of blind drives and auto-navigation software…. Sols 362 and 363 were planned together as another two-sol plan, again with the basic intent of driving as far as possible. After a directed drive of 90 meters (295 feet), the rover turned 180 degrees and continued in auto-navigation mode, resulting in an impressive 156.55-meter (513.6-foot) traverse. That is a new record for a single sol of driving on Mars.[32]

A later experiment demonstrated obstacle avoidance in a curved 15.8 m drive on sol 1,160 (April 29, 2007) using software uploaded to Spirit in 2006. The revised method combines "visual odometry to track the rover's actual position after each segment of the drive, avoidance of designated keep-out zones, and combining information from two sets of stereo images to consider a wide swath of terrain in analyzing the route."[33]

Despite years of experience as a roboticist, the working relationship Trebi-Ollennu developed over the years operating MER was unexpected: "It's a true partnership. That's what has surprised me on this project, that I picture the rover as my teammate *(laughs)*." Of course, the cooperative endeavor is asymmetric because the overall plan is crafted by Trebi-Ollennu through his knowledge of their relative capabilities: "I'm able to understand the limitations of the robot because of previous experience, part of the flight systems testing team. I know what it can do and what it cannot do, and I'm also able to know my limitations and that's very important." This relationship is like the symbiosis mentioned by Cabrol, here articulated as "my work" and what the rover as an external agent does— "your work." Rather than always being embodied in the rover, the rover planner projects a partner relation. Thus, Trebi-Ollennu says he "tells" the robot what to do and it can "think" for itself.[34]

Such human-controlled "partnership" between people and remote automation in space has its origins in the Surveyor project (1966–1968). Taking a detour for a moment to review Surveyor's operations provides a useful backdrop for better appreciating what has been accomplished on MER and the different ways in which robotic instruments and manipulators can be controlled (figure 6.9/plate 12).

Surveyor 3 was the first demonstration of remote control of a lander's arm, using tele-operation to the lunar surface. Following the successful engineering experience gained from Surveyor 5 and Surveyor 6, which carried a television camera and alpha-scattering instrument, Surveyor 7 (January 1968) was the first opportunity "for the scientists to call the shots." Like Surveyor 3, it carried a "soil mechanics sampler arm" but coupled this with an alpha-scattering instrument and television camera, which had to be operated in a novel way to accomplish the mission:

> After a look around at the "exciting" terrain, Turkevich's [alpha-backscatter] instrument was to be lowered to the surface, but it failed to drop. Roberson and his remote arm were brought into play, and gently lifted it to the surface. After one series of measurements, Roberson then dug a trench and moved the instrument to the freshly exposed soil at the bottom for another analysis. Finally, he lifted the instrument and placed it atop a rock. The sampler was also used to shade the instrument from the hot Sun. In addition to helping its scientific colleague, the sampler picked up and pushed clods, hit and weighed rocks, dug other trenches, and made more bearing tests. *The way the men, the camera, the arm, and the dry chemistry instrument worked as a team illustrated the power of a partnership.* This last mission put the frosting on a scientific expedition to Earth's nearest neighbor and was a turning point in automated spacecraft applications.[35]

The five Surveyors together provided seventeen months of lunar operations, which included "more than 87,000 photographs … and 6 chemical analyses of surface and subsur-face samples."[36] Although perhaps resembling MER in some ways, the fieldwork and time delay required a completely different exploration system for Mars. Geochemical analysis (with Mini-TES and APXS) plays a central role on MER, but these analyses are far more

Same   Different

Mobile Agents   Surveyor III/Apollo 12

Lunokhod-2   Cassini/Huygens

FIGURE 6.9/PLATE 12
Four ways in which people can interact with robotic systems, constituting radically different kinds of *exploration systems*, relating operations in time and place. Clockwise from lower right: Cassini-Huygens, like MER, involves operation at different places (Earth and Titan) at different times (Cassini is programmed for future operations; Huygens had a fixed program). Lunokhod, in contrast, was "joysticked," so human and robotic operations occurred in different places (Earth and Moon) at practically the same time (with the 1.4-second communications delay). Prototype experiments in NASA's Mobile Agents Project at the Mars Desert Research Station in Utah showed how people and robots can directly interact (same time, same place). Surveyor III and Apollo 12 astronauts operated at the same place on the moon at different times, illustrating the use of a robot for reconnaissance of human missions. *Image credits:* Clockwise from bottom right (all accessed February 14, 2008): Huygens/Cassini—top: NASA/JPL/USGS, colored radar image by Cassini, "Liquid Lakes on Titan," http://photojournal.jpl.nasa.gov/catalog/pia09102 (rotated and cropped); bottom: European Space Agency, "Artist's interpretation of Huygens on Titan," http://esamultimedia. esa.int/images/cassini_huygens/HuygensRocks01_H.jpg (cropped); Lunokhod—top: Moscow University, Department of Lunar and Planetary Research, Lunokhod-2 on Earth, http://selena.sai. msu.ru/Home/Spacecrafts/Lunokhod2/lunokhod2e.htm; bottom: Lunokhod-2 camera image, Don Mitchell, "Soviet Moon Images," http://www.mentallandscape.com/C_CatalogMoon.htm; Mobile Agents—photograph by the author, "Brent Garry and Abby Semple with Boudreaux EVA Robotic Assistant, near Hanksville, Utah," April 2005; see William J. Clancey et al., "Automating CapCom Using Mobile Agents and Robotic Assistants," *American Institute of Aeronautics and Astronautics 1st Space Exploration Conference*, January 31, 2005–February 1, 2005, Orlando, Florida, http:// eprints.aktors.org/375; Surveyor III/Apollo 12—Pete Conrad, taken by Al Bean during EVA-2 on November 20, 1969, http://www.hq.nasa.gov/alsj/a12/AS12-48-7133.jpg (cropped from top).

specific, frequent, and essential to ongoing decisions about what to do from day to day. MER's suite of interactive visualization, constraint checking, and program emulation tools could be only vaguely imagined in the 1960s (even graphic displays were not commonplace until the 1980s).

Reflecting the advances in computer automation, the science team's attention level has shifted from direct coordination of the components, which is now programmed and automated at the lowest levels, to real field exploration. Thus, the collaboration of the scientists among themselves and with the engineers shifts from the weeks required for the Viking mission to a matter of hours, and soon, perhaps, they will be talking about partnership with the planning system.

Squyres summarizes how MER's visualization tools fit into the overall concept of doing fieldwork through the rover, refining the point made earlier about providing tools a geologist would need (chapter 2):

> The visualization part of things is actually very, very important in what we do. Doing field geology is a process of assimilating information: scientific observations, geometric observations of the stuff around us, formulating a hypothesis, and then making an observation to test that hypothesis. I've devoted a lot of time and thought to taking what geologists actually do in the field and breaking it down into its component tasks—because we have to replicate all the tasks with a robot on another planet. I want to make sure that in the process of using a robot as a stand-in for a human explorer we don't lose something along the way. So you want to try to create a set of tools and procedures, operations, processes ... that enable a group of skilled geologists to accomplish everything that they could, that they would accomplish if they were there themselves, within the limitations of the set of tools that you have and the mobility capability.

It wasn't sufficient to bring along the tools a geologist would need; ways of seeing remotely were required to employ the instruments productively. Robotically mediated action is enabled by the visualization tools that orient the scientists and engineers by gluing together the data from Mars, what has been done so far, and tomorrow's commands (figures 1.2, 6.5, and 6.8). On Viking, the terrain around the lander could be mapped once and used throughout the mission (incorporating new trenches), though 3D color computer graphics was rare in the 1970s—an era when even photographs were not displayed on computer screens. Working with a rover, the terrain has to be mapped for each new location, and multiple locations were often visited by MER in a single week. Advances in computing made this possible, with rapid production of color 3D visualizations on what had become personal supercomputers.

In a pioneering effort, Stoker brought virtual reality tools to Pathfinder's operations that went beyond pasting together mosaics; she reasoned from field experience and by analogy to what was required for controlling underwater vehicles. She likened previous tools to "trying to construct a world model by looking at postage stamps." The Pathfinder science team became dependent on the MarsMap tool for pointing the camera; the software was further developed into the Viz program used for MER. Interviews by the Ames human-centered computing team with Stoker and others familiar with Pathfinder contributed to the awareness that the MER mission operations system had to provide university scientists with tools so that they could participate in planning the remote fieldwork—tools that heretofore were home-brewed, not integrated, and unsupported.[37]

During MER, scientists sometimes draw initial plans directly on photographs (figure 7.1), which can be useful in group brainstorming. However, great efficiency is gained by converting and combining data (e.g., the Mini-TES and Pancam overlays) into a virtual reality display, which enables assimilating what is present, understanding physical relationships, probing further to get more information and resolve causal questions, and generating parameterized commands for the instruments.[38] Including ten cameras on each rover (two Pancams, two Navcams, four Hazcams, plus the MI and descent imager [DIMES[39]]), melding these into graphic tools for locating the rover and instrument data, and simply seeing what is possible to study is the foundation for the experience of working on Mars.

In short, virtual reality is required for doing fieldwork with the sensing and manipulation tools MER provides, enabling people to observe and to look around (see figure 3.6). Perhaps the most substantial advantage is gained by aligning camera views to work with the instruments, then merging images with analytic measurements. The ability to see molecular composition (Mini-TES, Mössbauer, and APXS) begins to provide a cyborg twist to the experience (figure 6.10/plate 13), providing a capacity to perceive the landscape in new dimensions.

"Robotic Geologists"

In the discussion of field science, I related the sociohistorical and cognitive-engineering ways of talking about exploration, contrasting the metaphorical use of "explorer" with how exploring relates to scientific inquiry. In practice, the scientists experience "being the rover"—exactly the opposite of the "robotic geologist" metaphor, which leads to bizarre implications when taken literally: "The Voyager mission and its kind do not rely on human discoverers. The mantle of explorer rests on robots."[40] In operating robotic systems, planetary scientists and engineers are not confused about the nature of the machines; as simply stated by Nicks, "There were no such things as unmanned missions; it was merely a question of where man stood to conduct them."[41] This position fits very well with Trebi-Ollennu's explanation of partnership, which he explains in terms of the allocation and planning of the work. Following this line, I argue here that the third-person perspective, despite or perhaps because it obscures how the work is getting done, expresses an emotional experience that virtual presence allows and also facilitates the collaboration among the scientists that virtual presence using MER requires.

Origin and Value of the "Robotic Geologist" Shorthand

To understand how the robotic geologist terminology came about, a good place to begin is how the concept of the rover changed in the series of proposals, as captured by the project's evolving name: Mars Geologist Pathfinder, Mars Geologist Rover, Mars Mobile Pathfinder, Mars Geological Pathfinder, Mars Exploration Rover. These names show how the concept began as a "geologist" version of the 1997 Pathfinder lander (the stationary platform component for the tiny Sojourner rover), hence "Geologist Pathfinder." Then the instruments were emphasized as being onboard the rover itself, leading to the terms "Geologist Rover" and "Mobile Pathfinder." The emphasis on geology field science appears in the penultimate name, "Geological Pathfinder," which was generalized, perhaps in recognition of the other sciences involved, to the final name, "Mars Exploration Rover."[42]

Obviously, the robots are tools for geologists; simply put, MER is "a geologist's robot." Why do the words get turned around to "robotic geologist," referring to the tool as a person?

In some respects, the MER scientists can be poetic about the rover because robotic technology is not their area of scientific concern and therefore is not something they need to speak about in technical terms. In characterizing Mars, they are quite insistent about descriptive precision. They would object if someone called an outcrop a "rock wall" or the ripples of El Capitan "smiles." On sol 37, when someone in Spirit's soils science theme group mentioned that they were choosing an image of the day (for web posting) of a sandy area, someone else in the room said out loud, "Just don't call it dunes" (yet this usage persisted in the press; see the caption for figure 6.1).

In introducing the paradox of the "robotic geologist" phrase in chapter 1, I suggested that the scientists are appropriately focusing on the phenomena they study, not themselves or the engineering logistics of their work. Certainly, the team is capable of reducing the rover to being a "payload" and a list of operations when scientific demands require.[43] But something more is obviously involved in their articles and presentations, for the poetry goes beyond a mere gloss.

On reflection, it is amazing how well the parts fit together: aiming to tell the story of their work so that it will be understandable and interesting, they draw from a readily recognizable form—that the robot is like a person. They do this in a genuine and natural way, expressing their emotional attachment to rovers, and with a magnanimous flourish that arcs the scene of the expedition's historical importance. In this constructed view, they are not present, yet the spectacle of the rover deliberately highlights the difficulty and value of the team's efforts: "Spirit and Opportunity have been timid, easily frightened into immobility by small rocks…. But the new software should make them smarter and more courageous."[44] Read that again substituting "we" (the team) for the rover names—what is it about MER that makes this identification so easy and graceful?

Anthropomorphizing MER officially began with the selection of the names in the student essay competition; though "Thank you for the 'Spirit' and the 'Opportunity'"[45] suggests more the names of ships than people. The twin Viking landers in 1976 did not have nicknames, and we were content with the grand touring names Voyager and Pioneer. Decades later, with a shift in interest for "educational outreach," the Pathfinder rover, Sojourner (meaning "traveler"), was named through a student competition.[46] The Mars Science Laboratory has been dubbed Curiosity and described as having vital organs, a brain, a neck, a head, legs, and a means of "speaking" and "listening."[47] Perhaps this way of speaking about a machine is socially acceptable because it parallels how many people relate to pets. Recall that the field test model of MER was named FIDO.

Another clue about why people readily adopt metaphors for rovers is that no one calls Hubble a "robotic astronomer," suggesting that anthropomorphism requires the spacecraft to be at the site being studied (which for Hubble would entail being near distant stars, nebula, or galaxies) and sensing or manipulating physical materials under the scientists' control. (Of course, the Hubble telescope was already personified by its name. If we had called one of the MERs "Steve," the stories would have been different, too.) News releases about Cassini say that it "captures," "sees," "finds," and even "discovered," but the stories themselves focus on the physical phenomena (e.g., ripples in the rings, storms on Saturn, rain on Titan) and what lead scientists say about them.[48] The mission appears more as a collection of investigators, instruments, and wildly different and separated places. The team shares Cassini as a means of transport and remote sensing, rather than acting through it as a single body that touches and manipulates material.

Anthropomorphizing a rover is consistent with projecting oneself into its setting and actions, producing a sense of moving, probing, and looking in a coordinated way. This psychological experience makes statements like those in Squyres's eulogy understandable: "What's most remarkable to me about Spirit's mission is just how extensive her accomplishments became…. Spirit explored just as we would have, seeing a distant hill, climbing it, and showing us the vista from the summit. And she did it in a way that allowed everyone on Earth to be part of the adventure."[49]

Talking about Spirit in this manner is a form of generalization called *metonymy*, whereby a part—the rovers, the most visible object and focus of concern—represents the whole, that is, the mission itself, the exploration process, and the scientists' geological work.[50] Such language is common in the press; indeed, "the press" is itself an example of metonymy, where the original tool, a "printing press," now stands for the entire journalistic enterprise. Every day, we read and hear metonymy phrases, such as "The White House decided …" "Wall Street cheered …" and the ubiquitous "NASA today announced…."

Examples of metonymy abound in the scientists' speech. For example, Des Marais said, "Obviously, every instrument was going to write a paper," referring to the groups associated with MER instruments. The rover's name refers as well to the entire team's work, including the engineers: "Spirit continued to make progress on the rover's winter campaign of science observations."[51]

This shorthand is clear and practical, but it is perhaps more natural and socially acceptable within the mission because of its psychological truth—this way of talking reflects how the scientists project themselves into the rover in order to operate it. When they say, "The

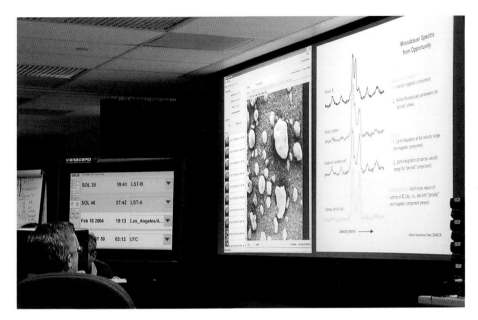

PLATE 1/FIGURE.1.1
Sol 25 in Meridiani science assessment room
end-of-sol meeting.

PLATE 2/FIGURE 1.2
One chosen placement for the rock abrasion
tool on El Capitan (sol 27).

PLATE 3/FIGURE 2.1
Map of Spirit's traverse from the base of the Columbia Hills to the summits.

PLATE 4/FIGURE 3.1
Geologist in Haughton Crater on Devon Island, July 1999.

PLATE 5/FIGURE 3.3
Samples collected by a single geologist in
Haughton Crater during one week.

PLATE 6/FIGURE 5.1
Athena Science Team and other MER
team members in early 2003.

PLATE 7/FIGURE 5.5
Gusev science assessment meeting.

PLATE 8/FIGURE 6.2
Simulated image of Opportunity superimposed
on Burns Cliff in Endurance crater.

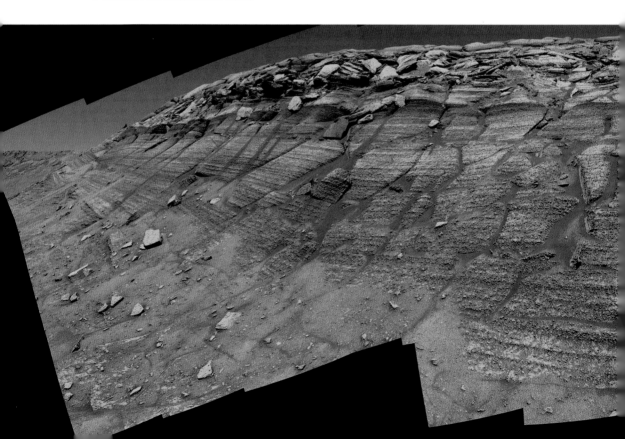

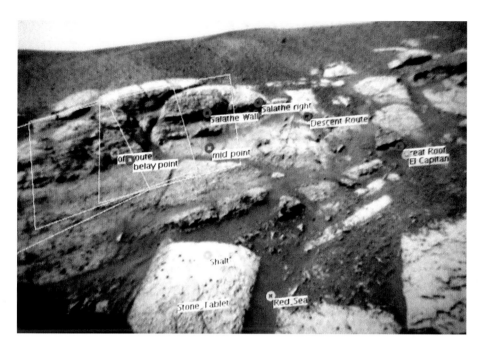

PLATE 9/FIGURE 6.5
Image overlaid in the Science Activity Planner
with target names and candidate images.

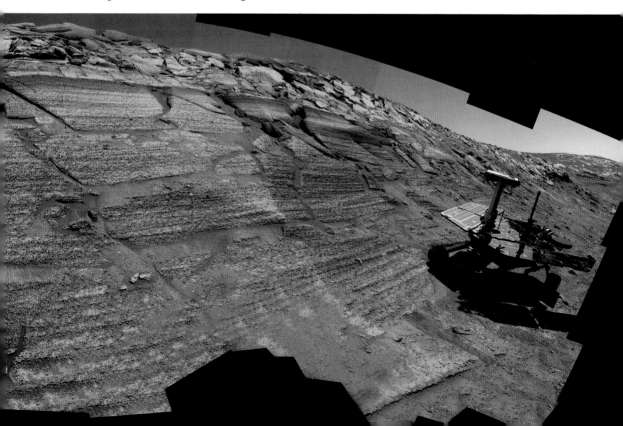

PLATE 10/FIGURE 6.6
Examining panorama of Eagle crater outcrop in
the science assessment room.

PLATE 11/FIGURE 6.7
Examining a 3D image in the Meridiani science
assessment room.

TIME

Same   Different

PLACE

Same

Mobile Agents

Surveyor III/Apollo 12

Different

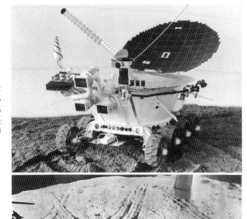

Lunokhod-2

Cassini/Huygens

PLATE 12/FIGURE 6.9
Four ways in which people can interact with
robotic systems, constituting different kinds
of *exploration systems*.

PLATE 13/FIGURE 6.10
Presenting Mini-TES data from Eagle crater
during science assessment meeting.

PLATE 14/FIGURE 7.2
Planning RAT and MI in El Capitan area of
Eagle crater.

PLATE 15/FIGURE 7.4
Atmosphere STG lead making observation
suggestions.

PLATE 16/ FIGURE 7.5
Rice preparing a presentation in the
science assessment room.

PLATE 17/FIGURE 8.1
Sims serving as Opportunity science
downlink coordinator.

PLATE 18/FIGURE 8.3
The Science Operations Working Group
(SOWG) meeting.

PLATE 19/FIGURE 8.4
SOWG meeting with engineers in the back
and science leads on the right.

PLATE 20/FIGURE 8.7
Scientists using SAP to plan RAT operations
and photographs in Eagle crater.

PLATE 21/FIGURE 8.8
Spirit at Hillary Outcrop on Husband Hill,
September 2005.

PLATE 22/FIGURE 9.2
NBC television crew filming the science assessment
room as Squyres presents.

PLATE 23/FIGURE 9.3
The Red Rover student team examining the
Eagle crater panorama.

PLATE 24/FIGURE 9.4
Spirit on the summit of Husband Hill.

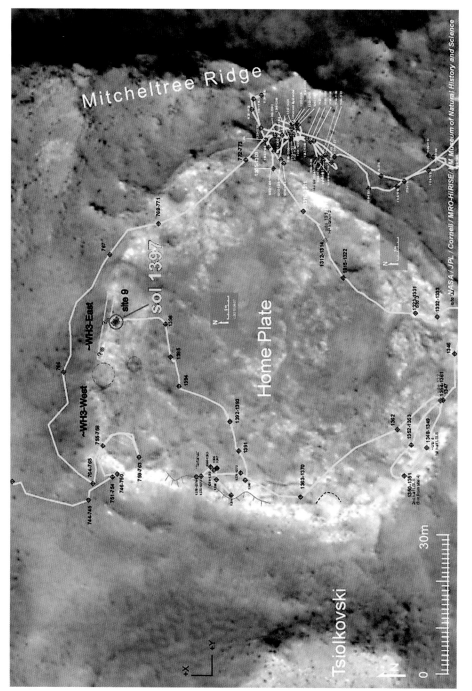

PLATE 25/FIGURE 10.1

Traverse map of Spirit in the Home Plate region behind the Columbia Hills.

rover is exploring," they say this because they know that they are exploring. This third-person form is written so convincingly and expressively (and perhaps sometimes obsessively) because they are speaking from the heart about themselves. The metaphor legitimizes an unexpected form of "clear speaking," one that hides the technical details of controlling the rover (fitting the scientific style) while revealing more clearly their struggles and accomplishments as a team (fitting their interest to tell a compelling story for the public).

The rover, as a visible and narrative focus of the mission's continuing saga, becomes a proxy for all of the work of the mission. JPL status reports (posted at first every sol, then about weekly) indicate where the rover is and what it is doing but only variously mention that a science team is even involved: "After receiving the results of the RAT brush, the science team will decide whether to look even deeper into the rock."[52] Questions of how they are analyzing data, their presentations to each other, and ongoing publishing are arguably not part of the "rover status" and are thus omitted.

From another perspective, the personification of the rovers encapsulates and reframes the loss of individual initiative and agency that working with a rover entails. Just as none of the scientists can today actually walk on Mars, scrape a rock, or take a photograph, no scientist can individually control the rover. The imbuing of the rover with agency reflects the indirectness and impersonal nature of everything the rover does. Only the engineers like Trebi-Ollennu can ever say, "I commanded the rover to do such and such." But even for the engineers, the decisions and operations are thoroughly social and indirect (e.g., figure 8.3).

So we have an inherent tension: unquestionable first-person imaginative projection by individuals into the rover's body and setting, but public narrative accounts that attribute actions to the rover, with unstated personal intents that are necessarily behind the rover's operations. Even when an individual or science theme group sways the team to get an observation onboard, the initiatives are not reported.

Yet paradoxically, for the work to get done, each scientist becomes the heroic adventurer in his or her own mind. Each scientist must place him or herself with "two boots on the ground," imagining what the rover can see and reach. And then, through a most amazing conceptual objectification of the whole process in which they struggle and debate with full awareness of the rover's clocklike workings, the scientists relegate themselves to behind the curtain as they say that the rover "sends postcards," becomes more courageous, digs a trench, and explores. No one scientist is allowed to take individual credit for discoveries on the mission, yet in these phrases some prideful self-description is projected onto the machine.

In summary, talking about the rovers as being like people is not a difficult stretch: like us, the rovers move, sense, scrape rock, move things around, take photographs, and send them home. For the MER scientists, being a member of a mission team realizes a personal dream of being an explorer. Yet the nature of telescience requires individuals trained in academic cultures that reward individual achievement and onstage performances to work as a group behind the scenes. Projecting their identities into the robots restores a sense of personal control, of being an agent on Mars, and helps make the anonymity acceptable. Everyone is equally unnamed yet equally present in the robot, moving and probing together. The result is an engaging story with an emotional connection. Viewed from afar, speaking of the rover as "discovering" may be inspiring or may have political and budgetary implications. For the project scientists, the phrase "robotic geologist" has practical personal and social value.[53]

Reconsidering Agency in MER and People

Although MER is not a scientist, is it an agent in any technical sense? The engineers have an experience related to that of the scientists: despite being aware of the scientific work and its meaning, they are predominantly focusing on a very different level of teleoperations, dealing with MER's health, computer system, communications, instrument behaviors, and movements. As related by Trebi-Ollennu's second-person narrative, MER arguably has agency in its "autonomous" operations, such as when it is driving long distances using only general location, time, and power constraints. Technically, one could call the rover a "navigation agent" or an "instrument deployment agent." Just don't call it a geologist.

So what distinguishes people as agents from certain kinds of programmed systems? MER differs from a human being by lacking an ability to conceptualize its actions in the context of social groups and values. For people, this conceptualization includes a socially grounded project and identity (e.g., knowing your role on the mission). This self-awareness is not merely an ability to recite and logically combine facts ("I am a robot on Mars; here is tosol's plan"), but a capability—by virtue of conceptually coordinating a conceived persona of motives, roles, and actions tied to emotional appraisals—to *participate* productively in a mission as conditions and opportunities change.[54]

Because poetic descriptions of spacecraft have been for so long expected in public and even historical presentations, the rover's lack of the human (and other animals') ability to conceptualize and to actually have experiences is usually entirely missed. For example, Squyres said in the fifth year of the mission, "They're less capable of exploring. They're less capable of improvising and responding to discoveries."[55] Of course, MER is not capable of doing any of these things at all (see figure 3.6). MER can find a preprogrammed feature (e.g., white rocks), but it cannot claim a formation to be new and interesting to the scientists.[56] If MER were improvising at all, the scientists would be struggling with much more than sand in getting it to go places.

Besides projecting themselves physically into the martian landscape, the MER scientists are capable of imaginatively bringing to bear their Earth experience as field scientists. They recognize in the martian landscape stories that feel familiar as they apply ways of seeing, segmenting, naming, and identifying the rocks and formations they encounter and find meaning by telling the causal story of the landscape. What the MER scientists see is literally "a landscape of possible interactions, rather than [just] an environment."[57] They are imagining future experiences, places to go, materials to inspect, and broader layers and horizons to relate to the story of Mars's climate and geological history. Without being able to interpret the horizon and to identify a part of the landscape as being potentially informative (e.g., Eagle outcrop, the Columbia Hills, a meteorite, the remains of a heat shield, almost pure white soil, the layers of Endeavour and Victoria craters, the distant rim of Endeavour), MER—like most of the engineers—is not capable of scientifically exploring Mars.

Besides constructing a narrative about Mars, the scientists are constructing a narrative about their work and themselves. Unlike a telerobot, "a monad with no history who 'behaves' in an objectively given world," the MER scientist is "an agent who carries on a narrative about herself in the world." MER operates only within the physical space of Mars; people operate within a conceptual space of roles, status, emotional values, and a sense of community, which defines a personal history with "directions for development." As one social scientist described this personal inquiry, "What is of interest to her is to follow complex flows of meaning relevant to the different choreographies in which she finds herself. Her representations and actions create her participation to such choreographies from moment to moment."[58]

In other words, MER scientists have persistent personal projects and interests, as well as trajectories of these, which they weave into the story of their participation, both in action and ongoing rationalization about their roles and contributions. The rover by contrast has no story, no ongoing personal narrative. Not just the autobiography is missing, but also the entire capability to have concerns and curiosity, which gets transformed into inquiry. Indeed, in our anthropomorphizing way, we make up a story, giving the rovers names, personalities, and experiences; viewing them as individuals with varying fortunes and difficulties; and saying, like a child's imaginary friend, that they are sending us postcards—because that's what someone important to us would do.[59]

What a Field Scientist Could Do in a Few Minutes

Finally, in this explication of "being the rover," we return to the not-always-joyful reality of doing field science on Mars with MER. The challenges are most visible in the scientists' complaints about the pace of the work, comparing distances to what they could cover. Squyres said, "It took four years to do a week's worth of fieldwork! It has unfolded in excruciatingly slow motion."[60] In the first two and half Earth years, the rovers together traveled about 16 km (10 miles). Considering that Mars has about the same land area as the planet Earth, this leaves a lot of unexplored terrain. Although exploration with Spirit was hampered by power and mobility, work with Opportunity continued, with detailed investigations of Victoria crater, for example, limiting travel to less than 2.5 km (1.5 miles) in the fifth year.

The scientists' complaints about speed makes explicit that they are relating MER to their fieldwork experience on Earth. For example, Carr said that one sol commanding (with data return sometimes delayed by days) was too slow, compared to the coordinated activity of seeing and acting in person: "The frustration is not so much not being there, not being physically and being able to hammer and everything. It's the slow turnaround time. Even though it's incredible what we are able to do and see, it's not fast enough. When you're in the field, it's seconds that you make the decisions, and here it's days. That's the frustrating part of it." Together with Squyres's related observation, "Humans have a way to deal with surprises, to improvise, to change their plans on the spot,"[61] we see that part of the pace problem is that the daily commanding cycle involves both too few operations and takes too long. Prepackaging an entire day of work feels too rigid and claims too much time; scientists repeatedly redirect their action as they look around, inspect, and move on. They don't think or work in terms of entire days. Fieldwork on Earth is more like wandering in a museum; MER is a group tour with a consensus itinerary—with each attraction requiring a day or more. Although the amount of data is substantial (given the kind and number of measurements and images), the process of learning about the place is in "slow motion."

Yet Carr and others realize that MER is doing a lot more than a field geologist: "the instrumentation is far more sophisticated." In part, the delays are caused by combining field and laboratory work. Working with this kind of planetary exploration system involves a fundamental contradiction that is difficult for everyone to grasp: we're on Mars, exploring the surface, but the rover is painfully slow—yet it's incredible what it does, and no geologist ever had such great tools in the field!

FIGURE 6.10/PLATE 13
Presenting Mini-TES data from Eagle crater during Meridiani science assessment meeting (February 11, 2004).

In talking through this contradiction, Des Marais articulates the oddity of the situation and concludes that the domains of experience are incommensurate. Asked to compare being in the field doing geology and working through a rover, Des Marais said: "It's an apple versus an orange thing; it's just so different. What a rover takes a day to do, we can do in about sixty seconds or forty-five seconds if we're out there ourselves. So from that perspective, why don't we get bored immediately?" Reminded that they're using instruments such as Mini-TES that scientists would not have used or been able to use in forty-five seconds in the field (figure 6.10) (or that anyone would be able to hold the MI precisely for a series of photos), Des Marais added, "Yes, and a key point is that some of these measurements take hours to make, and that just wouldn't work if you were out there doing geology. That would be a boring experience! *(laughs)* So that's a good example of why these are just qualitatively different domains of existence."

So in part, the pace must be slow to allow the instruments to be applied. A person could walk much faster, but without collecting the same data. The contradiction between their fieldwork experience and this Mars expedition stems mostly from their restricted mobility—despite being on Mars, they are moving across the landscape at best about 150 m (500 ft) in a day and are forced in the winter to stay in one place for a month or more. Working on Mars can feel tedious, as they must repeatedly wait for the robotic laboratory to go through its paces. Whether on Mars or on Earth, a scientist would not want to accompany a rover and wait for the instruments to operate. Working together with the robotic laboratory on Mars, the scientists would move ahead or send the rovers out in advance to provide reconnaissance photographs, molecular analyses, and perhaps some well-chosen micro-images.

Waiting out the winters is frustrating, too, for the rovers have much less power from the sun, and heating is more important, so the operating day is much shorter and drives are reduced or eliminated. Cabrol describes the situation of exploiting the rover yet keeping it safe as inherently conflicted:

> Sometimes you want to spend more time somewhere. And then you look at the clock and say, "If we say spend more time here, we're not making it to the point where it's the safe haven for winter, and then we're going to die." So what is the priority between not getting your data or losing all possibility of any future data? That was a sticking point with Home Plate because we are getting into a terrain that was so fascinating, so interesting. We are here with the most interesting stuff in Gusev, and we say, "Okay, drop the rock hammer and leave!" *(laughs)* But we are not far, so it means we still can come back.

What feels particularly odd when moving on Mars are the long traverses: distant targets are set, usually a crater or a hill, and the team heads for that location for months or even more than a year. In part, a drive can take longer because some scientists want to collect data about the rocks and soils they are passing. Comparing this again to personal experience in the field, Cabrol said, "The frustration I had goes back to human exploration. It took us six months to go from Columbia Memorial Station to Columbia Hills, whereas it would have taken half an hour for a geologist on foot in the field. But this is the way it is for now. It's not that it was imposed on us without us knowing it."

After arriving at the Columbia Hills, Spirit took about one Earth year to climb about 100 m from the base to the top (figure 2.4). Similarly, Opportunity traversed to Victoria crater for nearly an entire Mars year (687 Earth days, or about 669 sols). On such a path, every first encounter—an outcrop, a white rock—is exciting, but then discovering something new requires traveling, with longer distances between novel findings as experience accumulates. Also, many areas of Gusev and Meridiani are homogenous, with the same kinds of soils, rocks, and morphologies, so there is a sense that less is happening ("less science return"). Overall, surveying and getting to know the martian landscape with the rovers is a slow process.

Furthermore, each traverse is ever forward, only rarely looping back to check something—a strange way to explore a landscape. Des Marais explains that on Earth we survey an area and then revisit sites:

> It's very important in the field, and you don't even think about it, that you can go back to some place. In standard field geology, if you're going to work an area one day, you run around and see what's there, and have a lunch and cogitate on what you've seen. Then you say, "I'd really like to look at these things more and get samples from here and there." And that's what you do during the afternoon. A fundamental dynamic of field geology is that you do a preview, and then you go back and cherry-pick what you saw.

Therefore, some of the frustration might stem from geologists' proclivity to climb up and survey before doing detailed studies, plus a preference for iterating over the same landscape. Rice explained his experience similarly: "Being a field geologist, you recon the whole area, and then you come back. In areas of greatest importance, you come back and hit it with everything you've got, your instrument suite or your samples, whatever's your field geology method." The field practice using MER is forced to be different because of the rover's relative lack of mobility.

Furthermore, on Mars, with a ninety-day warranty for the rovers, one had to treat each sol as the last and move on to new things. Des Marais explains: "Squyres before we landed was even saying, 'Well you know, we're going to have a sniper mentality here, where you think the mission's going to end any minute.' There are so many things that can go wrong. So you're just going to want to get into this mentality of grab what you can." Yingst joined the mission long after the warranty was up and uses the same expression: "It is very much like *(laughs)* doing geology with a sniper aiming at you." This makes the scientists "anxious to get certain kinds of data that we have not gotten yet."

Des Marais contrasts this urgency with the pull exerted by an understanding of proper field methods. He thought, "We really need to get ourselves more into a mode of where we can survey and decide what we want to pay more attention to. Well, that turned out to be totally unrealistic." Instead of the field science practice of "survey and return," the rover

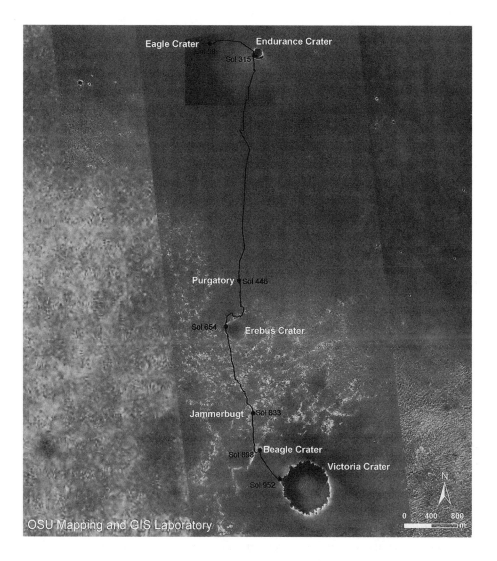

FIGURE 6.11
How scientists moved Opportunity from landing
spot (at top) to Victoria crater (sol 955). *Image
credit:* JPL/OSU Mapping and GIS Laboratory.
Source: JPL MER Mission, Detailed Traverse
Maps, October 1, 2006, http://marsrovers.jpl.
nasa.gov/mission/tm-opportunity/opportunity-
sol955.html (accessed January 1, 2008).

is driven over a strikingly direct path (figure 6.11). Des Marais explains how analysis that occurred after moving on once did lead the Spirit team to "go backward," which revealed their "ever forward" mentality:

> The reality is that everyday that we do something, we're probably not going to do that again. In fact, going back, the rover actually driving back to where it was, became sort of like an anathema to us. When we actually did do that on Cumberland Ridge and went back to that white soil, all the people on the web were saying things like, "Oh, my god, they're going backward—what's happening!" *(laughs)* Well, it was a compelling enough thing that we had digested, we said, "I want to go back and take another bite." *(laughs)* That was a rare event. I think we've done it maybe two or three times.

Important exceptions to the "no going back" rule are the traverses near Home Plate behind the Columbia Hills to return to a slope for the winter to maximize the solar power and hence the necessary heating of the rover's instruments and mechanical parts. Also, in July 2009, Opportunity was turned back about 180 m (590 ft) to the Block Island meteorite, which the team did not notice until the rover had trundled past it.[62] Reflecting on the difficulty of keeping the rover moving to the next destination in a manner informed by the ongoing analysis, Des Marais says:

> On Earth you do things in stages: you do the fieldwork and then analyses; after that, you start putting your thoughts together and writing a paper. You compartmentalize your work that way. Whereas on MER, you're forced every day or few days to do a much more thorough analysis of your current situation. You go almost all the way through the traditional scientific process on Earth with every few days of experience you have on Mars, because you've gone to the next place. It's like a chess game—you're sitting there thinking out all the permutations, what it is to be where you are, because you're not coming back 99 percent of the time. Even though the rover only does what we could do on the Earth in the field in a minute, that's a gross oversimplification of how much we're actually doing in a day.

Here Des Marais has articulated how much extra work is really going on. He recognizes that the rover's day involves tasks that a geologist in the field would never do, let alone complete in a few minutes. Furthermore, the collaborative practice of managing the rover so that it can apply the instruments involves a great deal of effort that a small team of specialists in the field—even if provided with a robotic laboratory with the same instruments—could not accomplish themselves. This work involves not only the programming of the instruments (such as the calibration with photos that Yingst mentioned), but also the decisions of which instruments to apply where and what the results mean.

Given the general shared frustration over the pace, how did the team decide when to stay and study an area more and when to move on? Staying or moving is a scientific and practical issue that was handled in different ways at different places for both Spirit and Opportunity. As a rule, the team completed their investigation of an area before moving on. The strategy has been to be thorough, and to not allow the thought of what lay over the horizon to tempt hurrying.

As an LTP lead, one of Des Marais's principal tasks was "to achieve and maintain a balance between detailed study of the site at hand and the need to 'use the wheels' to broaden the field of investigation." Perception and judgment changed as the investigation progressed. After characterizing the most commonly occurring features at the landing sites, the balance moved from detailed studies toward a more concerted effort to explore new sites.[63] The interest of course is not just in being in new places, but finding something different, the overall objective being to identify and understand the diversity of materials and their topographic relations within the region.

Particularly revealing is the multiple-year trek to Endeavour crater, announced with a fanfare in the September 2008 JPL MER report "Road Trip Gets Under Way." Rather than minimizing the time required to arrive at the destination, the scientists stopped repeatedly to investigate dozens of rocks and outcrops and several significant craters. They were not merely moving to Endeavour but always conducting field science, which involves inter-rupting the journey to characterize diverse materials and morphologies. They made haste between such stops to get to Endeavour, as the relatively uniform (and hence familiar), unconsolidated soils of the Meridiani plains often allowed.

Remarkably, the expedition paused over three months at Santa Maria crater (December 2010 to late March 2011). About a month of this time can be attributed to the solar conjunc-tion, which inhibited communications between MER and Earth. The 23 February 2011 mis-sion report a week after the conjunction said that departure was imminent, but the team didn't leave for another month, clearly revealing their unhurried approach. Sims explained:

> Santa Maria was a planned campaign long before arrival. It included two long baseline stereos and it definitely took longer than we planned: it is a crater with interesting orbital mineralogy and of a similar size to Endurance, so we wanted it as a comparative data point. This was an investigation with a science team member who was a strong advocate (Wes Watters), which is usually critical to something being fully developed. We were committed to doing it "right" within limits! Every few days, we weighed the option of continuing observations or continuing traverse. This is typical of our longer observations.[64]

Sims summarized the method of focusing versus getting to see as much as possible as "good (or maybe better) science trumps the game." The journey continued, and two months later, about 3 km on the odometer after Santa Maria, another target beckoned: "The sci-ence team has spied an outcrop ahead to perform some brief in situ (contact) science."[65] Thirty kilometers from Opportunity's landing site in Eagle crater, it's difficult not to smile at their repeated reminder, "We won't stay too long."

Ironically, the pace of the exploration enables the communal activity—it provides time for people to put their heads together and reason through what they are seeing and how best to use the rover. The result is a kind of "collective human brain" controlling the rover. Reflecting on this situation, Squyres mentions the oddity of being in a group, the slow pace, and then the fantastic collaboration that results:

We're doing geologic fieldwork. Normally, when you do geologic fieldwork, you do so with one partner, maybe two—a small group of people out in the field, working together, traveling light, moving fast, trying to get science done. You're making observations, making decisions about where to walk next, what measurements to make, and what rocks to collect and so forth on a very quick basis in a field setting. *Here,* you've got this mission that works at an absolutely *glacial* pace. Oh, god! It's so tedious sometimes! It takes *forever* for this rover to do anything—it drives me crazy.

At the same time, you look at what we've accomplished in 1,400 sols and it's dazzling. But on a day-by-day basis, it's 3 meters and a cloud of dust! But the interesting thing about the fact that it moves so slowly is that you can actually take the assembled brain power of dozens and dozens of highly trained, highly capable people—if you can figure out a mechanism to get all of these prima donnas working together—you can harness that assembled brain power to do a much better job of making decisions and making observations and formulating plans and executing those plans and analyzing the data than you could ever do if it was just two or three of you out in the field together. And so the glacial pace is actually *enabling* being able to get this huge team of scientists to work together.

Squyres reminds us that the emergent overall quality of the work is important; the pace has a positive effect that transcends personal frustrations. Here the synergy of the exploration system comes into play (see figure 4.3)—continuing engagement over 1,400 sols would be less assured if the turnaround cycle were measured in weeks as on Viking, or if the team didn't operate by consensus, or if looking, naming, moving, and probing weren't so well integrated into "one instrument."

Also, field science is not a race: in evaluating the MER exploration system, we must consider what is being learned about geochemistry, mineralogy, geomorphology, climate, and so on. The Apollo astronauts driving rovers also "covered more ground in less time" than MER,[66] but they were not analyzing samples (even with a hand lens) as they moved along. It is tempting to evaluate telerobotic methods with simple metrics such as speed, the number of photographs, weight of samples, and others. But scientific value is measured by evidence for questions of interest and discoveries promoting theoretical insight. Ultimately, comparisons of exploration systems must be post hoc, for we can't properly evaluate MER's hardware, software, or operations strategies until we know more about what was beyond the next hill (e.g., on Von Braun Planitia) that might have dramatically changed our understanding of Mars and/or how we investigate the planet. For this reason, "mission success" criteria are defined as narrow operational metrics (e.g., number of targets, instruments applied, distance driven, mission duration) rather than scientific objectives. In interpreting the scientists' complaints, we are trying to understand what scientific objectives could have been better satisfied. In part, we need to know what the scientists couldn't reach or were unable to analyze—the unknown unknowns, what they might have discovered but missed (a point I take up in the concluding chapter in considering future missions).

Felt experience of pace can also be put in historical perspective. Field science on Earth has not always proceeded as fast as explorers might like. Referring to nineteenth-century fieldwork, Rudwick comments that the "slow rate of travel … forced the geologist to accept a certain leisureliness of experience … time to reflect on the significance of what he had already seen." In some respects, this may be true for the daily commanding cycle on MER, but the reported "glacial" rate of travel from one major feature to another, as well as the forced linear approach among the disciplines (e.g., waiting a few days to do a detailed analysis of a single rock), confines the geologists to a particular spot amid many potentially interesting places to go and study. Thus individuals may experience being physically stuck and held back, which is quite different from "waiting for the wind to change and for the ship to sail" or "walking through an interesting landscape beside a donkey."[67]

As the rovers aged and the budget dwindled, a donkey might have seemed preferable. By late 2007, the scientists scheduled end-of-sol meetings (to review which of their proposals translated into concrete rover plans) once a week, rather than daily. Squyres said that this schedule was sufficient because "the pace of discovery is less frantic than it used to be." Some of the methods to save energy are to work on one sol then charge batteries the next and skip a daily data transmission or two. These changes lower costs, too, supporting requests for continued mission extensions.[68]

In summary, the scientists' frustration arises because of the cognitive dissonance between imagining presence on Mars and the time required to get information about the area they are exploring—learning about small areas like Home Plate stretches over many months and even years. These delays make the mission more like an eighteenth-century voyage (albeit with a full complement of specialists onboard, not one or two polymaths). The scientists become like shipmates bound for a distant land they crave to know more about. Committed to years of working together, they survey the landscape, drawing maps and sharing ideas to build a model of the geomorphology and climatological evolution of the region. And slowly, they progress, moving more or less straight ahead toward a distant and unknown but hopeful place. Early on, after arriving at Home Plate, Rice dubbed Spirit's next, ultimately unreachable target to the southwest "the Promised Land."

A Consuming Activity

So, after all, what is it like to be working on Mars? The flip side of the scientists' realization of a lifelong dream is the partial nightmare of the nominal mission's day-for-night, seven-day schedule, followed by years of often daily participation in telecons. In an hour-long interview, Sims used the term "consuming" five times. As an engineer (computer scientist) reflecting on the personal demands of MER operations, he opined:

> MER was a disaster on those things on my view. We survived. We managed…. But the model in MER that we had is all about a *consuming vehicle* for a set of people. You have a set of people—105 or 130—for which this is a consuming activity. What I would actually like to do is have one person who can control 100 robots on the moon—we run telescopes that way. It's not a crazy or unreasonable idea. And we have the technology to almost do that. That's a very different operations point of view than MER. On MER, we spend lots of hours checking everybody else's work and sequence development.

So in Sims's opinion, the technology, the organization, and the consensus process required unnecessarily complex, interactive teamwork, which wears people out.

Jim Bell spoke similarly about the fatigue, long after returning to Cornell: "We have become engrossed with and consumed by Mars. Maybe that's why I can't remember what a normal day job used to be like!"[69] Carr describes the personal cost of participation during the nominal mission:

> The SOWG chairs came in late in the science day, then went through the engineering. Well, I wanted to overlap more with the science, you know? So I'd come in two hours or so before, and of course I got really tired. My memory of being there, during the engineering part of the day, is that I was tired all the time. I was just exhausted. It's not just the switching of the time, but the length of time that I was at the lab. I was there because I was interested in science! Yet I had to go through the whole process, until the command load was sent. I was there for twelve to fourteen hours. The rotation was four days. I would go and walk around the lab to wake myself up, to keep myself awake.

Carr wonders whether age accounted for his fatigue. But Rice was thirty-four years younger and had a similar experience, possibly because he pushed himself correspondingly harder:

> I was there all the time. I never would sleep. No exaggerating. I was living on four hours of sleep in two two-hour shifts. The way I looked at it, this was a once in a lifetime opportunity to work with two rovers on Mars. And the ninety days kept hanging over our heads.
>
> I thought: "Ninety days! Why not go for it? When would I get a chance to do this ever again?" Imagine … it is going to be a while before it happens again, if ever. I was selected to be there, so I'm going to live this thing to the fullest extent possible. I don't regret doing any of it. I look back now and I wonder how I did it. But the human body adapts pretty quickly to weird situations. Even with the lack of sleep, there's always excitement and just the pure sense of exploration that kept me going.

In most respects, the MER exploration system is designed for such participation. The technologies available—affordable telecons lasting a hour or two; web-based tools for posting and sharing plans, data, and analyses; email; graphic presentations—sustained the team's communications and awareness of the rovers' progress. MER invited contribution and commitment. The tools and the individual scientists' motivation led them to persist, to work harder, and to fit this activity into other responsibilities and missions.

Different technology mentioned by Sims, some of which is already available, would change participation and hence an individual's sense of working with a rover on Mars (although it might not improve the quality of the work). So one of the aspects of MER we must examine in more detail is the communal nature of exploration that the MER methods have brought about. An engineer might talk about one person *controlling* a hundred robots, but a more realistic problem is enabling a hundred scientists to participate in controlling a single robot. What facilitates MER's consensus approach and what is it like to be part of it?

7

The Communal Scientist

An inherent aspect of "being the rover" is that it can be in only one place at a time, so sometimes excruciating decisions must be made about a particular path and the inquiry that everyone must follow. The scientists move together across the plains of Mars, looking and probing the rocks as one body. As the mind of the rover, they merge their observations into one agreed plan. Rice explained that the decision in early 2006 about whether to go clockwise or counterclockwise around Home Plate (figure 10.1)—whether to see the south end, seeking new clues for understanding this feature, or whether to go to a known safe place for the winter—was "probably the most wrenching and detailed discussion in Gusev the whole mission." Their situation might seem at times humorously absurd, if it weren't so necessary, tedious, and significant. In this first expedition on Mars, they are all in the same boat, and survival comes first.

The rover technology, safety concerns, and other public and economic constraints of the mission effectively transform the scientific process of exploring Mars to conform to textbook methods of doing science that are driven by systematic hypotheses fitting shared interests. Inherent disciplinary conflicts can also arise—for example, between geochemists and geologists—because laboratory and field scientists have different data gathering practices. Most of the scientists are used to working alone or in small groups, which are not always multidisciplinary, but on the mission they are obligated to a communal endeavor. Nevertheless, by adapting their methods and learning from each other, scientists and engineers alike have developed what the team members describe as "the best collaboration."

Although people often refer to all kinds of joint activity as being "collaboration," we can distinguish different ways of working together in a mission like MER. "Collaboration" implies a shared *intellectual project* (e.g., deciding how to investigate Victoria crater), whereas "teamwork" might be just be a cooperative *physical* activity (e.g., manipulating FIDO into a truck). "Communal" suggests a more pervasive surrendering of personal space and choice about scheduling and activities. Viewed from the perspective of the group, communal activity develops holistically, both cognitively in multidisciplinary goals and weaving of methods and emotionally in adopting a responsible attitude and empathy for others' ideas and needs. Objectifying this communal aspect during the nominal mission, each rover team lived and worked together in their own martian time zone and privileged rooms at JPL, with the floors color-coded to designate the rover and theme group name tags around the science assessment room to label people by their role.

How did this kind of collaboration come about? How do individuals experience the communal activity with its consensus approach, which is so different from being in charge of their own investigations in the lab or field?

One Instrument, One Team

As mentioned in discussing MER's relation to other missions, Squyres's role, and telerobotic inquiry, the Athena team conceived the rover holistically with instruments that fit together in their operation and together covered what a field scientist would want to do on Mars. But it remained to develop roles and processes that fit the rover's design.

To begin, consider how the pace of the mission enabled so many people to contribute to the daily plans and the ongoing long-term trajectory of the rovers. On the one hand, shifting the planning cycle from sixteen to twenty days on Viking to one sol on MER (see figure 4.1) sustains engagement. On the other hand, a daily turnaround, as opposed to joy-sticking from Phobos or reprogramming multiple times during the day, allows more people (living in different time zones on Earth) to contribute to the interpretation of data and ongoing planning. Besides the external logistics of an organization, facilities, work processes, and tools, in some way the right attitude had to be constructed as well. The MER exploration system had to be designed so that people worked together, which Squyres refers to as "running a happy ship":

> Now, having the operations take place at a slow enough pace is necessary, but not sufficient, to harness all that brainpower. The other thing that you have to do, as a PI, is create a set of operations processes and create a working environment. I don't mean a building with desks and computers, but a mindset among the team that is highly collaborative. And you've got to create what people in the Navy talk about as "running a happy ship." It's not an easy thing to do. Trying to create an environment in which people work together well, respect one another, don't get on one another's nerves—after four years—that's tough. You've got to keep it fun. If it isn't fun, people are actually going to get tired and walk away from it. So one of the things I tried really hard to do was create a working environment on the team in which everybody feels like their opinions matter, that their views are listened to. No, we don't do what everybody wants. You can't, because there are too many people and too many different ideas. But everybody feels like they're a valued member of the team, and it's still fun to come to work in the morning.[1]

The challenge was made more difficult because Squyres did not personally choose every member of the science team. As noted in chapter 5 (table 5.1), Squyres handpicked the co-investigators on the proposal, and when a second rover was added, NASA doubled the team through peer-reviewed selection. Squyres then had to construct an integrated "total team" by identifying appropriate roles and responsibilities in the science operations process: "Who were the good SOWG chairs? Who were the good long-term planning leads? Who were the good documentarians, and so forth. I still select and schedule all those people. It's been a real interesting exercise in the psychology of getting a bunch of smart people to work together, as a team."

The objective is clear, but how is it actually accomplished with scientists? People with PhDs are trained to develop individual, well-reasoned perspectives. During the mission, for example, an interest group emerged called "The Dust Devil Advocates."[2] How do the Science Theme Groups mitigate against this fragmentation into cliques and ownership arguments?

> This is a fundamental point. This is one of the most important things about how this mission was organized, and it's one of the things I'm proudest about as PI. It would have been very

easy, very natural, to structure the team by science instrument. Look at MSL: the natural structure is that there's a Mastcam team, and there's a SAM team, and there's a ChemMin team, and a ChemCam team, and so forth. We don't have a Mini-TES team, and a Pancam team, and an APXS team, and a Mössbauer team, and an MI team. Now we do have individuals or groups whose responsibility is to do the downlink task and to do the uplink task for those instruments. But the rank-and-file scientists are not assigned to instruments. And that's fundamental.

The whole idea behind this payload [the MER instrument package] is that you have a suite of tools that work together in a complementary fashion. And you don't want to have the Pancam guys arguing with the Mini-TES guys, arguing with the APXS guys about the availability of resources: "I want to use Pancam today! I want to use Mini-TES today." And they fight with each other. Instead, what I always told everybody is that *we have one instrument*—it's the rover. It's got a bunch of sensors that work together in a complementary fashion…. I want to hear geologists arguing with geochemists.

Did Squyres say all of this to the scientists? "Well, I didn't quite tell them that. I just *structured* the team such that it would work out that way." Comparing to Pathfinder, we can see how operating more instruments creates both an engineering problem of managing the programming constraints and an organizational problem of managing the scientist's individual interests, knowledge, and opinions. Again, in contrast, on Viking when programming limitations precluded daily commanding, scientists had less involvement; for example, although instrument groups could make suggestions, "the decision about what pictures to take was made by the imaging team alone."[3]

Many people (including space scientists and engineers) might not understand, even after years of operating Spirit and Opportunity, how fundamentally different a rover mission is from operating an interplanetary spacecraft. Instead of a collection of instruments that independently collect a variety of data, such as the six fields-and-particles instruments on Galileo, the MER team engages in a collaborative process of selecting and studying well-targeted locations by moving all of the instruments together from place to place. On Cassini, different subteams studied the rings and moons and essentially owned and operated the spacecraft at different times and places in the orbits of Saturn. In contrast, the MER scientists are investigating the same targets on a surface landscape in a cross-disciplinary way—they must not only agree on where the platform will be pointed and whose instruments will be operating; they are going places and doing things together. Their consensus is at the level of what is being investigated and how the instruments are applied to resolve scientific questions, not just managing shared resources at the engineering, pragmatic level of operating the spacecraft.

Comparing a planetary flyby or orbital mission to a rover mission like MER involves making a distinction between *cooperation*—sharing common resources (e.g., power, time, bandwidth, the payload platform)—and *collaboration*—engaging in a common inquiry from different scientific perspectives. For example, Galileo had seven PIs for the pod of atmospheric probe instruments, six PIs for the six fields-and-particles instruments, and four PIs for the remote sensing instruments.[4] Generally speaking, Galileo's PIs were cooperating, not collaborating.

Reflecting the topographic regions of the Saturn system, the Cassini mission has three operations groups that focus on different geographic areas of inquiry: Titan, Saturn, and the icy moons. They share the spacecraft resources and time by partitioning orbit segments and then engage in "collective planning."[5] Without a scan platform, they must turn the entire spacecraft, and thus can generally use only one instrument at a time, leading to a "complex trade system." Trajectories may be planned up to a year in advance, making route planning very different from an overland expedition like MER. However, because Cassini is powered by a plutonium radioisotope thermoelectric generator (RTG), it can operate all the time; the nominal mission was four years. Some opportunistic retargeting is possible, such as flying through the plumes of Enceladus multiple times in 2008. Nevertheless, the scientists' engagement on Cassini is at the opposite end of the spectrum from MER's daily contingent planning. Cassini's long waits—imposed by physics, the geographically defined investigations, and the spacecraft's design—engender less sense of working together as a single "Cassini Project Team" than MER, requiring the patience of waiting in turn (besides waiting to get somewhere) and ultimately causing more frustration and questions about lack of fairness.[6]

Cassini's different instrument teams necessarily must cooperate, although they have different goals and interests. This kind of resource-coordinated, independent inquiry occurs on MER, too. For example, MER atmosphere studies are generally independent of geochemistry studies. But MER's instrument teams also often collaborate when deciding what features to investigate. Constrained by a common geography, they are exploring Mars together. So this is an important lesson, too: the topography and regions being explored affects the design of the payload and platform; these and the science planning tools will determine what forms of joint activity are desirable and possible.

Reflecting the new kind of mission—conducting overland field science—Squyres started with a simple, logical concept of the rover as a single, coherently operating entity: the "robotic geologist." This concept plays out in the selection of the instruments, the systems engineering to make sure they work together, and the visualization tools. The holistic concept of the rover (contrasted with a robotic platform carrying a "payload" of independently operated instruments) flows naturally into the design of the organization for collaboration—one instrument, one team:

> It was just intuitively obvious to me given the payload that we had put together. You've got these sensors and each of them provides complementary bits of knowledge, so that the totality is more than the sum of the individual parts. *You're going to use the payload to fullest advantage, if people look at it as being entirely at their disposal.* So what happens is that I'll have geologists or geochemists come in to the meeting and say, "Well, we can really understand this if we first take a Pancam of it, hit it with Mini-TES, and if it looks like this, go over and APXS it." You know, that's the idea! The whole idea behind MER is that these tools work together. Look at the silica discovery. Okay, the mobility system, which

we use as a soil physical processes tool, trenches up some soil. We notice it with Pancam; we hit with Mini-TES. It looks interesting, and we go over and we figure out what it's made of with APXS. Everything works together.

Indeed, MER fits together on many levels in several ways: sensors fit sensors, the rover fits a field geologist, and disciplinary contributions fit each other. So rather than "my instrument," we have "my shared rover"?

Exactly! And so it was just really evident to me from the start that if you structured the team along *those* lines that you have geologists arguing with mineralogists arguing with atmospheric scientists. Well, that's what scientists do, naturally. If you were out there in the field, doing geology with you and your field partner, you might be arguing about what this rock means or what that rock means, but you're not going to argue about, "Well, should we use the rock hammer or should we use the Brunton compass?"

But given the originality of this concept and its variation from the instrument-centric organization of other missions, didn't he get any pushback? Squyres says, "I'm not sure that they even noticed that I pulled it off on them. *(laughs)* No, I didn't get any pushback. And it's worked." This unification of interests helped people share the rover for different interests, avoiding the plight of individual instrument teams feeling like they are being slighted: "You use the instruments when it's time to use the instruments, in the service of the science you're trying to do. It's a pretty fundamental point about how the mission's set up."

This transparent "one instrument, one team" discipline has a special manifestation in the decision-making process itself, as the investigation must also be a team effort, requiring actions to be justified in relation to each other and systematized to endure the scrutiny of many other scientists for years to come. This is a fundamental paradox of being a scientific explorer on Mars working remotely through a rover: the scientists have to work as if they are writing a textbook in the field.

Being a Textbook Scientist

Every sol, scientists in the theme groups proposed "observations" for the rover to perform. The term "observation" emphasizes how the scientists project themselves into the rover's actions, but also testifies to the level of abstraction at which they plan the rover's behaviors, allowing them to think in terms of what they wanted to know. Observations involve applying instruments, perhaps in multiple, different ways, and always to particular "targets." As presented in chapter 1, Squyres clearly articulated the rules of the game, the need to be driven by hypotheses, as they began the first ratting in Eagle crater. In practice, requiring scientists to articulate a hypothesis that an observation was intended to test had many fundamental purposes: sharing a limited resource with competing disciplinary interests, the possibility that the mission could end without warning at any time, the expense of each observation,[7] and the highly visible and historic work, requiring careful documentation of motivations and context.

Asked about his remarks about stating hypotheses during the Meridiani science assessment meeting on sol 25, Squyres explains that working with a rover changes the nature of geology and field science, citing the unknown longevity of the rover, the public nature of the work, and the cost:

> It's a different style of geology, it's a different style of doing field science. You can't screw around because your vehicle could die at any time; you've got CNN watching your every move; *hundreds of millions of dollars were spent*. You know, it's not like I just got dropped off at my field site, and we'll kick some rocks around for eight hours, and go into town and have a beer, and come out again the next day and, yeah, if we did it wrong, we'll do it right tomorrow. It's a different style of science. You have to take your responsibilities extremely seriously.

These responsibilities are demanding in new ways, so the scientists' behavior must adapt correspondingly. Once again, what Squyres called the hyperdimensional[8] requirements that bound the design and operations of the rover into "one instrument, one team" can be serendipitously exploited to bring the scientists together:

> Another thing is that if you take this "let's try this; let's try that" approach, it makes it a lot harder to get twenty or thirty or fifty scientists all on the same page. Because if you've got twenty geologists, you're probably going to have twenty-one different ideas for what you ought to do that day. *(laughs)* Whereas, if you say, "All right, the decisions we make operationally have to be hypothesis-based … "—in other words, come to an SOWG meeting to advocate a specific measurement, but don't say, "I think this might be interesting, let's give it a try." Tell me *why* it might be interesting. Tell me what specific hypothesis you can test, what the outcome of that hypothesis might be. Crazy hypotheses are welcome!

Here the constraint of making the mission public was qualified to enable brainstorming during the science sessions. So that people wouldn't feel inhibited, they didn't allow the news media into the end-of-sol meetings. Combining this privacy to be creative with critical thinking binds the group further, as they work like one mind with a hundred sources of good ideas:

> It's almost like you have a real rapid turn around form of peer review. Your ideas have to stand up to the scrutiny of your colleagues, who have ideas of their own. You try to foster an environment in which ideas of all kinds are encouraged and respected, but in the end, the best ideas win. If you can do that, then instead of 100 people with an IQ of a 120, it's like having one guy with an IQ of 10,000.

Throughout, the method of deliberately expressing hypotheses forced the individuals to better formulate their thoughts, but also promoted learning from each other. What did the other scientists think of the textbook approach? Carr recognized the objectives and advantages:

> It did seem somewhat artificial, but it had a purpose. When you have limited resources, you have got to say, "Well, is it better to drill here or drill there? We can't do both of them." And then you get started talking about the merits. Ultimately, one person chooses one

place over another because he has in the back of his mind a hypothesis, a suspicion, that this is going to yield this, whereas the other person has another hypothesis. When you're in the field, you don't do that. But I think the formalization of the hypothesis was a healthy thing to do.

Rice explained further that beyond the unknown duration of the mission, the RAT in particular had a limited life:

The one on Spirit is no longer usable [as of mid-2005, after grinding fifteen targets[9]]; we ate the diamond heads out of it. I think that was the reason for articulating hypotheses. "Let's not just go off ratting everything in creation here," because we may need the RAT down the road. We're going to need it in Victoria crater now to see if there's different rock down there. It's a shame we don't have it on Spirit right now [in August 2006].

Because they couldn't practically go back to previously visited places and they were pressed for time, scientists were sometimes forced to make difficult decisions about how to respond to an individual scientist's initiative. A striking case of balancing time and return occurred early in Opportunity's mission. After the outcrop found in Eagle crater had been ratted, Grotzinger argued for taking ten sols to study the outcrop in detail (figure 7.1 and figure 7.2/plate 14). Squyres exclaimed that they couldn't spend more than 10 percent of the entire mission on a fishing expedition and required a shorter campaign.[10] Three sols later, Grotzinger had cut it down to imaging just two rocks, and the result was one of the most important finds of the entire mission: evidence that rock layers had been deposited by water.

In general, individuals were thus required to articulate to the group a scientific reason for going out of the way, for staying an extra sol or more, for selecting a given rock for applying a given instrument. Early in the mission, the scientists couldn't be sure if Spirit would survive the nearly 3 km journey to the Columbia Hills. As a compromise, the team developed a plan to acquire systematic data (called "ground truth"—an unwitting geo-logical pun) that would be useful for calibrating orbital observations. This plan addressed the legitimate concerns of the scientists and instrument teams that the "forced march" might prevent Spirit from acquiring very much science data before she died. A moving survey was the default or fallback plan at this early stage of the mission, with key observations repeated every fourth sol in what Des Marais dubbed "the sol quartet."

To formalize the team's rationalization of the daily plan, each observation is documented by a "hypothesis" entry in a template (table 7.1). Cabrol explained that the approach forced upon the team is something like what a professor "would ask a student to do in the field, which is basically a guided outline of research, not serendipitous." This data-gathering strategy is a commonly articulated principle in teaching. Applying it to MER made the investigation systematic and deliberative, not merely opportunistic and reactive, which fieldwork on Earth usually allows.

FIGURE 7.1
Markings indicating areas of interest and RAT candidates of Eagle crater outcrop, presented as a large panorama photography on the table in the Meridiani science assessment room (February 19, 2004). See the group examining and further marking this photograph in figure 7.2.

FIGURE 7.2/PLATE 14
John Grotzinger, Jim Rice, Brad Jolliff, and Bob Anderson in MER Meridiani science assessment room, February 20, 2004, after the science context meeting on sol 27, planning RAT and MI in El Capitan area of Eagle crater. The panorama was printed approximately lifesized. (These are four digital video frames.)

TABLE 7.1

Sol 15 Opportunity (February 7, 2004) Constraint Editor Activity Report details. Following the summary page (figure 7.3), the purpose of the observation, specific methods, and constraints are recorded, as well as (optionally) the scientific hypothesis.

OBS. #	NAME	PURPOSE	SCIENTIFIC HYPOTHESIS
1	IDD STONE MTN PIEDMONT (SOIL)	EXAMINE MORPHOLOGY AND COMPOSITION OF STONE MTN	—
2	IDD STONE MTN ROBERT E (SOIL)	EXAMINE MORPHOLOGY AND COMPOSITION OF STONE MTN	—
3	PANCAM AFTERNOON LEFTOF STONE MTN (GEO)	PANCAM MINERALOGY SIGNATURES	—
4	PANCAM BLUEBERRY HILL (GEO)	EXAMINE MORPHOLOGY	—
5	PANCAM NOON SKY COLUMN (ATM)	QUANTIFY SKY BRIGHTNESS AT NOON TO BASELINE SOLAR ARRAY PERFORMANCE	SKY BRIGHTNESS AFFECTS SOLAR ARRAYS
6	PANCAM AM PARACHUTE L456 (PIO)	POSSIBLE DUST DEVILS	—
7	PANCAM TAU ANYTIME (ATM)	QUANTIFY ATMOSPHERIC OPTICAL DEPTH IN TWO CHANNELS	ATMOSPHERIC AEROSOLS VARY WITH TIME
8	PANCAM TAU ANYTIME (ATM)	" "	" "
9	PANCAM TAU ANYTIME (ATM)	" "	" "
10	PANCAM TAU ANYTIME (ATM)	" "	" "
11	PANCAM TAU ANYTIME (ATM)	" "	" "
12	NAVCAM FOR DRIVE POINTING (CHEM.)	ACQUIRE NAVCAM TO CONFIRM POINTING AND FOR DRIVE TARGETS	--

Des Marais explains that the rigor of systematic study requires and should be shaped by an understanding of different hypotheses, but this also depends on already having a broad model of the overall terrain and historical processes that shaped it:

> There are two scales of the activity that should be kept in mind here. On the broader, macro scale of doing science, serendipity and the technology that happens to be available are contingency aspects that work against the textbook ideal of developing a hypothesis and so forth. But I think at a micro scale, where maybe you're in a situation where you have defined the situation to some degree, there are only so many things you can do. There's a way you should probably do your observations to control for all the variables. Because you have the chance, it's really most effective to drop into a more rigid process of hypothesis testing, with a defined set of observations. I think that's where the team found themselves at that point in Eagle crater.

Des Marais recounts the same story told by Rice and Squyres, emphasizing the momentous difficulty for the scientists:

> The very contentious thing was when Grotzinger came back and said, "You know the way to really do this in a way that approaches the rigor that we would do this on the Earth, we need so many sols, and so many resources." And Squyres said, "This is ridiculous, we can't do this, this is too much." And Grotzinger sort of stumbled back out of the room, "God, what am I going to do? Because I'm caught between what I know is essential, versus what's apparently going to be possible." And of course the rest is history; it obviously worked out.

Reflecting further on their choices in Eagle crater and the trek to the Columbia Hills, Squyres comments that many decisions were with an eye to an unknown time horizon, and that knowing more, you'd sometimes make different decisions:

> You've got to find this balance between the long view and the immediate view. That's something that we have struggled with from day one. I'll be honest with you, if I had known that these rovers were going to last 1,400 sols—if the day we landed you said, "We're still going to be doing flight ops in December of 2007," there were things I would have done differently. The list is endless. Certainly more time in some places. We would have spent more time at Eagle crater. I cracked the whip on those guys, I said, "Sol 60 we're going over the lip—I don't care." I said that about sol 45: "You've got two weeks, that's it, then we're out of here." I made myself very unpopular with the team, but I felt that we had to do that. I wouldn't have laid down that dictum if I'd had *known* that we were going to survive for 1,400 sols. But at that point, you know, the engineers said, "It's a ninety-sol mission." And you've got to treat that seriously.

It is easy to forget after years of continued operation how worry affects people early in a short-duration mission. In June 2008, I observed Phoenix mission operations for ten days. After thirty sols, some of the scientists' concern about meeting the mission objectives within the ninety-day nominal mission bordered on anxiety. At first, this seemed dramatically different from our experience on MER, until I remembered the tension in studying Eagle crater.

The constraints and choices, especially early on, are overwhelming to balance and sort out into a plan: differing disciplinary interests, opportunity (the visible, beckoning outcrop versus the greater unknowns beyond the crater), the mission objectives that define success (distance, images, instrument applications), and the uncertainty of the equipment's longevity. In that perspective, the decision to allow two-thirds of the nominal mission (sixty out of ninety sols) to be spent exploring the first spot spied in Eagle crater was a serious commitment, scientifically well reasoned, but also daring.

Yingst provided a broad, philosophical, and historical perspective of the careful textbook approach that framed these difficult decisions. She locates the group's exploration in their shared methodical training and broader responsibility to the community and the future:

> I think you have to have a strong scientific rationale if you're going to use a multimillion-dollar machine to look at something. You're using a very limited resource and you're also *doing geology for the ages. (spoken with reverence)* We talked about this just this morning. One of the things that drives the process is the idea that this is the only data that we're going to get from Gusev or Meridiani probably for a very long time. I did my doctoral thesis primarily on images by Lunar Orbiter from the 1960s! This is going to be a huge gold mine for many years to come for many geologists, both terrestrial and non. We have a big responsibility to make sure that the data set is as complete but also as compelling as possible.

As one might expect, the hypothesis-driven discipline for making observations does not necessarily carry over to the practice of presenting data to the team for consideration. Here an open, noncommittal, and informal approach is often appropriate. For example, I observed many times during the nominal mission that scientists would show and point to features without interpreting them. On M25, Rice was viewing images by himself on a laptop in the long-term planning area throughout the science assessment and end-of-sol meetings. Periodically, people came over to talk to him about an image that he had left visible on his screen for the longest time, a spherule with an apparent scoop cut into it. In this end-of-sol meeting, he took his laptop to the front and connected it to a projector. He showed a half-dozen images as "something to think about—just a curiosity." He stated no hypotheses, and gave no interpretations, just pointing as one might in the field: "Hmm, look at this."

Of course, very broad hypotheses guide overall MER science and are always of interest to refine—Mars once had flowing water; there may have been niches for life; and life may still exist somewhere. Such hypotheses are implicit in many observations and most obvious in published justification for the choice of instruments[11] and how the data are analyzed (e.g., comparing prevalence of selected compounds in a selected subset of rocks).

Aside from the discipline of articulating hypotheses in planning discussions, the scientists were encouraged to document hypotheses in their written plans. Based on Pathfinder experience, Golombek promoted a kind of "experiments notebook … to place data in a conceptual framework so someone could delve into it later."[12] Similarly, based on Viking experience, Arvidson emphasized how easy it was to lose the reason for taking certain types of pictures or selecting certain targets. The idea was to give future analysts "a picture and story, not just a picture."[13] Accordingly, MER's "documentarian" (a member of the LTP group) kept notes of science meetings, recording key interpretations, hypotheses, planning decisions, results, and the many constraints affecting the work.

Furthermore, to capture this context at the point of defining an instrument application, a template was devised for formalizing Observation proposals, including information to be filled out about objectives called "intent frames." To understand how these Observation intent frames were actually used, I analyzed two preliminary (nonfinal) sequence plans from Gusev and Opportunity on 7 February 2004, totaling twenty-eight observations: one drive, ten Pancams for the Atmosphere science theme group, two APXS, four Mini-TES, two Navcams, and nine other Pancams. For these twenty-eight observations, the Constraint Editor Activity Report listed fifteen hypotheses, of which nine are identical: "Atmospheric aerosols vary with time." Table 7.1 summarizes the observations and documented purposes and hypotheses from sol 15 for Opportunity. For comparison, figure 7.3 shows the first page of a high-level summary of the plan for that sol.

Constraint Editor Activity Report
Sol 15 / Opportunity (MER-B, in Meridiani)
Generated on February 7, 2004 at 20:04:05 UTC

Resource Summary

	Duration (HH:MM:SS)	Energy (Wh)	Critical bits	Non-critical bits
Total	20:05:54	134.2	5.1M	76.5M
Priority 0	18:05:17	89.0	4.1M	20.8M
Priority 1	01:46:20	40.4	1021.3k	54.1M
Priority 2	00:03:05	1.0	21.1k	1.2M
Priority 3	00:11:12	3.7	42.2k	457.1k

Observation/Activity Summary

Crit?	Prl.	Name	Duration (HH:MM:SS)	Energy (Wh)	Crit Vol. (bits)	NCrit Vol. (bits)
	1	IDD Stone Mtn piedmont (soil)	00:22:35	12.5	244.9k	23.8M
	0	IDD Unstow [none] IDD_UNSTOW	00:03:45	2.0	117.2k	1.2M
	0	VERIFY MI PLACEMENT [p1111] HAZCAM_FRONT	00:00:49	0.2	10.6k	512.0k
	0	4bpp Piedmont [d0010_p2939] MI	00:18:01	10.3	117.2k	22.1M
!	0	IDD Stone Mtn Robert E (soil)	17:59:46	87.3	2.0M	17.7M
	1	VERIFY MI PLACEMENT [p1111] HAZCAM_FRONT	00:00:49	0.2	10.6k	512.0k
	0	left 4bpp Robert E [d0010_p2933] MI	00:18:01	10.3	117.2k	6.1M
	0	right 4bpp Robert E [d0010_p2933??] MI	00:18:01	10.3	117.2k	6.1M
!	0	Robert E [d0010] APXS	05:09:19	20.6	375.4k	2.1M
	2	VERIFY APXS PLACEMENT [p1121] HAZCAM_FRONT	00:00:49	0.2	10.6k	512.0k
!	1	Robert E [d0011_n1502_n1921] MB	12:11:59	45.5	1.4M	2.1M

FIGURE 7.3
Constraint Editor Activity Report, a high-level specification for the sequence of Opportunity operations on sol 15.

The Constraint Editor Activity Report is generated after the SOWG meeting, as part of the sequence generation process. At the top is a summary of the resources required (the primary constraints) for prioritized groups of operations: time, energy, and data return (in critical and noncritical amounts). The Observation/Activity Summary lists the operations in chronological order. The first Activity involves three "observations": unstowing the arm (IDD), photographing the placement of the MI with the Hazcam, and making an MI photograph. The target is a soil area on the top of Stone Mountain, part of the outcrop in Eagle crater. The second observation uses the MI, APXS, and MB on another soil area of Stone Mountain, called "Robert E."

Even when the hypothesis slot of the intent frame is filled, the scientists often provided information about data characteristics, not the causal interpretations. Another slot, called "Purpose," was always filled in. The entry was sometimes general, such as "examine morphology and composition" (table 7.1), but more often it just stated what data the instrument would acquire: "Stereo mosaic of Robert E.," "Document placement of MI," "Determine dust and cloud opacity." Notably, the aerosol "hypothesis" is not being tested by this observation; it has been previously established on earth and is not in dispute. The objective is to *document how aerosols are varying here*, in this place on Mars, at this site at this time of year.

By my interpretation, only two of the documented "hypotheses" were indeed hypotheses being tested by the indicated observation (e.g., a hypothesis about what the Pancam images might show: "The bounce marks will brighten with time due to dust fall out"). One of the "hypotheses" is gathering data for engineering ("Checking driving accuracy"). The documented purposes of the thirteen observations for which no hypothesis are specified indicate that they are part of systematic efforts to characterize materials or areas. The terms used are: examine, quantify, long-term monitoring, "for routine study," "to characterize," "to acquire compositional knowledge," "to observe," and "systematic observation."

As a further test, I examined the sol 12 Meridiani Activity Report. It specifies twenty-five observations, many of which are similar to those planned for Gusev. None of the observations specified hypotheses to be tested. In summary, 2/53 or fewer than 4 percent of the observations for Opportunity on sols 12 and 15 and Spirit on sol 37 were documented as testing hypotheses.

From this analysis and the discussions in the science meetings, we can conclude that talking about hypotheses is important for prioritizing and allocating the rover as a resource among dozens of scientists with competing interests. Articulating hypotheses educates the entire team about each person's thinking and helps them move the exploration forward together in a systematic, responsible way. For example, the decisions about which rocks to approach and study and how long to spend in a particular area, as well as the example emphasized here—how and where to RAT an area—are more strategic and require reasoned arguments. Moving to the target, verifying placement of the IDD, ratting, and then perhaps micro-imaging the result might require two or even three sols, making each RAT targeting a costly decision compounded by concern about the limited life of the abrasive surface. We find highly detailed records of science goals, summaries of hypotheses and evidence, and planned observations in memos by individual scientists (e.g., "Completing Meridiani Outcrop Science") and the daily Documentarian reports.

However, when it gets down to specifying instrument work in an already chosen area over a period of time already agreed upon, many of the documented Observations are just that: simply gathering data, like a field scientist surveying a new terrain. Even when specific interpretations might be under consideration (e.g., the outcrop is layered, the layers are caused by water sedimentation), the measurements to test these hypotheses are usually independent of the interpretation (e.g., if the hypothesis were layering caused by wind-blown sand, the same data would be desired). This generality—where the data will help disambiguate multiple interpretations, including the current hypothesis—could explain why the "scientific hypothesis" slot was omitted from the Observation intent frame, but strategic "campaign" planning is well articulated in the Documentarian's notes.[14]

Field versus Lab Scientists: Disciplinary Conflict from Pace

Although the MER science team is plainly a harmonious, extremely successful group, this collaboration has developed over time and involved dealing with different interests and priorities. Examining the issues carefully, we find that the distinctions that arise on Cassini over geographic interests arise more specifically on MER over different scientific perspectives and methods for investigating the surface—roughly as materials versus the morphologies of a topography—complicated by scientists who are looking into the sky and might not care what is underfoot (figure 7.4/plate 15).

FIGURE 7.4/PLATE15
The Atmosphere STG lead, Mike Wolff, makes observation suggestions in the science assessment room (February 11, 2004).

As mentioned, when Spirit set out on the long traverse toward the Columbia Hills, Des Marais satisfied disciplinary interests by suggesting the sol quartet. When I observed the science assessment meetings in February 2004, before this scheme was implemented, the Atmosphere science theme group appeared to be on the short end of the stick, sol after sol. Because their measurements usually involved pointing a camera into the sky, other scientists could dismiss their observation proposals as something that could be done anywhere at anytime.

An unusually tense interaction occurred during the Gusev SOWG meeting on G37 as the final plan for the next sol was taking shape. On seeing that his group's request was going to be cut, an STG lead said in a flat tone, "The other STG guys want to go to bed, so we will eliminate our requests entirely." He had said earlier that they specified highest priority because their observation was omitted the day before. He was asked by the SOWG chair to explain the method and its importance. He said that omitting part of it was unacceptable; you might just as well delete it all. So that's what happened. He seemed politely but visibly upset. Although these discussions were always collegial, fatigue—easily attributable to living on Mars time for over a month—and delays did have an effect.

Even with the sol quartet, an inherent conflict remained about how far to drive each sol; this decision directly affected how far apart instrument operations would occur, which was important to the geochemists interested in a systematic survey. The conflict between the geologists and other scientists on this point was rather humorous, at least in retrospect, as Carr recounts:

> I was mostly involved in Spirit and I was very concerned about pace early on, because we landed on this basaltic plain. We thought, "Well maybe if go to up to the [Bonneville] crater rim here *(gestures)* and we look in the crater *(gestures)*, we'll see something more interesting." We went up there, and we didn't. We kind of realized then that this landing site was not going to address the problems that we went there for, the problems of water. Very early on, I thought, "Jeez, we need to get to some more interesting place." So there developed a conflict between the geologists and the chemists; we wanted to get out of there. But the chemists never saw soil or a rock or anything that they didn't want to analyze! Because they have these instruments there, they want to use them. And they want to stop at every rock! And they were just basalts! They were basalts, you know! *(laughs)* We wanted to get to the hills and see if there was anything different in the hills, because that looked to be our best chance of getting into something that was not just basalt, this boring basaltic plain we were on.

Carr also noted that the SOWG chair had discretion for resolving constraints that emerged late in the planning process. The policy of planning by consensus wasn't always possible, so in the morning an individual might be unpleasantly surprised:

> There was a geochemical measurement that I cut out of the plan, because it was either that or move. I thought the integration time on the chemical experiment was short anyway, so the data was going to be marginal. Well, when the scientist found out that I had cut his observation, he got really, really pissed off. And it was part of this problem of the pace. Some of us had had enough of basalt. And we wanted to get to the hills. And sure enough, when we got to hills, everything was different!

As daily commanding raised expectations for everyone's regular participation, it may have also increased frustration. Objectives, opportunities, and resources were not always aligned. Carr's strain from the urgency and competing interests is palpable: "We didn't know then that we were going to have a lot of time in the hills. We saw the clock ticking down—we had a ninety-day mission. And here we were, messing around, analyzing common-variety basalts that had no weathering or anything."

In delving more deeply, we find that the conflict of pace and sampling between the geologists and chemists more generally reflects the difference between laboratory and field practices. Kohler's study, *Landscapes and Labscapes*, cautions how scientists attempting to combine methods lead "troubled lives": "It is one thing to combine two laboratory disciplines, like chemistry and experimental biology, and quite another to live comfortably while straddling the lab-field border."[15]

Accordingly, MER field scientists—predominately the geologists, who were engaged in physically exploring the landscape as they do on Earth—were particularly frustrated by the delays caused by the laboratory-type analysis. For example, Rice explains that from the moment he saw the hills, he wanted to take off: "I was bringing it up every chance in discussion with the group. I said, 'This is great, but really, we have to get to the hills. There's gold in them thar hills.'" Rice complained to Squyres, who was then SOWG chair, that the group wasn't exploring Gusev as a geologist would:

> I came to Squyres once because I got frustrated in Gusev. I thought we weren't going anywhere. It took us awhile to get to Bonneville crater, looking at every little sand grain and every little baby rock out there in the field. I mean, there's more to the story than just what's in front of our eyes here…. You don't sit there when you first get to a field site, at just that one spot. You climb the highest hill, survey the landscape, get a feel for the land, and walk around, get an idea what's going on. To me we weren't doing field geology early on; we weren't going anywhere.

On G36 during this period, Rice said to me after a science assessment meeting, "I'm not a petrologist; I'm a geomorphologist." In other words, "exploring scientifically" means different things to different disciplines: "You could see early on who was a field geologist and who wasn't in the group. A field geologist saw the value in driving and looking around, in surveying the land. Whereas a lot of people just wanted to sit there and analyze every little sand grain." The technology enabled much more than what they were doing: "These things were rovers, they weren't landers. They were designed to cover terrain."

Professional interests affect what you want to look at and therefore how you move about. With the geologists and chemists all piled onboard Spirit, heading for the Columbia Hills together at an average 42 m/day, some scientists were thrilled by the nearby attractions, while others stayed below deck waiting to get to their destination. And that wait could be tedious, as the rover's typical daily movement could be walked in about thirty seconds. But getting off and walking was not an option. So on G36, Rice sat in the science assessment room reading a paper about Apollo 11, which he hoped would provide some historical perspective.

Without a doubt, the scientists respect each other and understand that there's a rationale behind other people's work. What these frustrations reflect is the pressing need to be responsible to one's own discipline and to carry out the investigation appropriately. But Rice's remarks reveal also that how to investigate Mars—that is, the allocation of time and instruments—reflected differing interpretations of how "the scientific method" should be realized at this time and place with these tools. With its "one instrument, one team" design, MER forged an unusual, communal field investigation that at times required fundamental compromises.

A simple view might be that the group consisted of field scientists and laboratory ("theoretical") scientists. But the population is highly varied, and experience doesn't divide easily by discipline. In particular, many scientists on the MER team had no prior field experience, including some planetary geologists (e.g., lunar or Mars photogeologists). Perhaps a third of the scientists develop theoretical models on the basis of fundamental laws (e.g., theoretical geochemists applying chemical thermodynamics to geology) or on the basis of traditional experiments. For example, Scott McLennan's group from SUNY Buffalo performed laboratory experiments to measure the effects of weathering (e.g., temperature, atmosphere, fluid composition) on synthetic basalts. In general technical terms, they described their interests as "evaluating the evolution of planetary crusts and surficial [e.g., evaporative] processes using the chemical composition of sedimentary rocks."[16]

Furthermore, instrument scientists who developed instruments and/or applied instruments may combine a wide variety of field, experimental, and engineering expertise. For example, Phil Christensen, codeveloper of the Mini-TES (also PI for the 2001 Mars Odyssey Thermal Emission Imaging System [THEMIS] instrument, and the Thermal Emission Spectrometer [TES] instrument on Mars Global Surveyor [MGS]), described his research as using "infrared spectroscopy, radiometry, laboratory spectroscopic measurements, field observations, and numerical modeling, in a wide range of field sites."[17]

Similarly, Darby Dyar, an astronomer who applied the Mössbauer Spectrometer, "uses several different types of spectroscopy to study rocks that originated from 90- to 0-km depth in the Earth, as well as lunar rocks and Martian meteorite samples collected from Antarctica." The Mössbauer was developed in Germany in the 1950s and was used by laboratory geologists to characterize Apollo's rock samples. Dyar's laboratory work provides a calibrated "key" for interpreting the instrument's readings at the –200°F temperatures on Mars.[18]

As the work proceeded over the years, Dyar and other scientists returned to their labs to do new physical analyses, checking instrument sensitivities by systematically testing response to known entities. For example, Yingst found it useful to create a library of terrestrial analog images at the hand lens scale for comparison to MER's microscopic images.[19] Such shifting between lab and field is forging new crossover identities, fitting Kohler's observations: "In a patchy cultural landscape, isolation and chance opportunities for mixing—or introgression—make possible novel kinds of practice."[20] Christensen, Dyar, and

Yingst are just a few of the many MER scientists who are adapting tools, ways of analyzing data, and field methods to the constraints and affordances of operating a programmed laboratory on Mars.

In short, for everyone on the team, the combination of scientific exploration in the field, using the instruments and analyzing data while on traverse was new. Some planetary scientists were on a field expedition for the first time. Many of the scientists in the geo-chemistry/mineralogy, soil and rocks physical properties, and atmosphere theme groups were gathering data systematically on long traverses for the first time. And the instrument developers were embedded in a scientific collaboration—discovering how people used their tools, how they played with other instruments, and how to organize and correlate data from different sites and formations over multiple years. In Kohler's terms, doing field science on MER is inherently "in the border zone" for everyone because they are working in both the field and laboratory arenas simultaneously.

As members of the MER project, the scientists' focus is on understanding Mars and contributing to the team, and as a rule they find a way to fit—they have been selected in part for this reason. However, from the subteams' and individual scientists' perspectives, further tensions may occur within their laboratory or disciplinary community. Exploring further the influences of external identities is beyond what we can cover here. However, we can be relatively sure that the interests and concerns that scientists raise during MER operations reflect pressures from the values and practices of their institutional and academic communities.

With this background now in mind, let's look again at how people viewed and dealt with the field-laboratory perspectives on the way to the Columbia Hills. Squyres recognized the inherent conflict of how to use the rover collectively; he bracketed the opposing perspectives by setting sol 160 as the goal for arriving at the Columbia Hills. Those wanting more time for science argued for sol 170. Rice and others who thought "we were wasting our time on the plains" argued for sol 150.[21] But, as Rice relates, Squyres also encouraged him to keep the destination visible to the group:

> When I went to Squyres and voiced my frustration with Spirit, he said something like, "I know, believe me, it's going to change." I said, "Steve, you still want me to be the cattle prod to get us to Columbia Hills?" "Yes, keep on bringing it up, keep on doing it." I said, "All right, I'll keep on doing it 'cause I think it's the most important thing we ought to be doing in Gusev crater." So I kept bringing it up every chance.

Sitting in that darkened room sol after sol, Rice felt as if he were on Mars and being physically held back by Spirit's slow pace: "I was kind of antsy, and sure enough, they started clamping down. Ray Arvidson [then the SOWG chair] said, 'We've got to get to Bonneville crater. Here's our target date.' People started being forced to do things like that."

Rice's role continued in a similar way in the years that followed, as he looked at an enticing area only about 800 m away from Home Plate: "I've been that way about the

Promised Land for two years now, too. I'm really going to be frustrated and really have a lot of regrets if we don't make it there. If for some reason we can't get there, the mission ends or whatever, that's just going to be something that is always going to be on our minds, 'What is that? The lakebeds?!' Who knows?"

Des Marais has often been in charge of long-term planning for Gusev and, as the results attest, feels satisfied with the decisions made during the run to the Columbia Hills:

> My job on Spirit is to coordinate between several different disciplines. The big elephant in the tent here are all the operational constraints. Let's say we had known that the Columbia Hills were sweet. I think it would have been a rush to judgment to just beeline to it. I think the whole idea behind the four-sol quartet was not just to keep people in the room from killing each other *(laughs)*, but also because there really was value in making a set of measurements as we went across those plains.

Like Squyres, Des Marais had a responsibility of framing the exploration from a broad perspective and, like Squyres, views the opposing views as defining a "negotiating space" that as a manager he works within:

> People like Rice are valuable. Maybe I'm a political moderate here, and so this is why I'm in this position. You need those forces pulling, to occupy the center ground. The instrument guys, who would want to use their instrument every day, don't want to move. Rice wants to move. So now you've got this negotiating space created by these people pulling each other. All you have to do is sit there and tweak the balance that develops in the group discussion, and you've got yourself a plan.

Yingst mentions that a scientist needs to know how and when to speak up, to judge the difference between what's important and perhaps uniquely known to herself, and what's only of interest to her:

> So as frustrated as I get sometimes, thinking, "We're still digging in the dirt!" and "I'd really like to look at some of these rocks over here," I also understand the compelling nature of the science that isn't in my field of expertise. I am there because I have a specific expertise, and I am there to defend it vigorously—that's my job. But I also understand that I'm going to be overruled probably more times than not because there are so many other pressing needs. And when my need is the most compelling, then that's what we do; and when it's not, then I totally understand in the larger sense why I step back. We do what is most important, not just for our team, not just for the scientists who are currently working on Mars, but for the scientists 150 years from now who are going to work on Mars.

The individual scientist may experience his or her perspective as being different and must check with others to establish whether the viewpoint must be preserved and how strongly it should be articulated. Correspondingly, leaders encouraged and valued individuals who spoke up. As Yingst says, it is their job. But everyone understands that they are studying a planet and gathering data that will be used for years to come by many other scientists. Thus, different personally conceived destinations, local targets, and timings were held in balance and assembled into a fair, rationalized, and historically responsible plan. Opinions were aired in the meetings and backrooms; the daily and long-term actions were negotiated. So finally, the group has one story to tell, related publicly as the exploits of the communal rover and what the scientific community learned about Mars.

Designed Collaboration: The Science Theme Groups

In considering the consensus approach to working with the rover, it is useful to recall that the practice could have been very different. Each instrument could have had a principal investigator, who could have made decisions about that instrument. Alternatively, universities or other institutions could have had control of instruments for certain periods, submitting proposals for observations to be made to an independent organization, as for Hubble, which peer-reviews and schedules the telescope's operations.[22] Individuals or disciplinary groups on the science team could also have had possession of a rover for a sol or a week, or even taken over control in some region.

Although instrument ("payload") specialists on any mission must always find a way to share resources, a rover poses additional constraints because the groups must decide where to go, and this choice can be more opportunistic than for orbiters and flybys. Nevertheless, broad temporal and regional agreements are useful. Rather than allowing groups or individuals to wander about, requiring drives to get back to where others want to go, the traverses are broadly planned into *campaigns* (tables 2.2 and 2.3) on an approximate schedule, with many local choices about pictures and analyses accounting for disciplinary interests.

As previously described, Squyres designed the team for collaboration by organizing the scientists into "science theme groups," with the LTP group and SOWG chair negotiating differences. (The concept of the SOWG is derived from the EOWG on Pathfinder,[23] but with only the APXS onboard, a theme-group structure was not required.) The MER investigation is notable for monthly rotating roles, clear leadership, strong partnership with engineering, informal collaboration between disciplines, and written "rules of the road" (policies for data sharing and publication).

Squyres believes that having the team physically work together at JPL, so they could get to know each other and establish decision-making practices, was essential for the later success in a distributed, telecon-based mode:

> There is absolutely no substitute for working together side by side in the same room for months at the start and getting to know each other. You cannot go immediately to distributed operations. [Now] I know my team inside and out. I know the engineers. I know these people so well I can listen to the meetings over the speakerphone, and I can hear the stress in a rover planner's voice, and know I'm pushing a little too hard on this drive, it's time to back off a little bit. This camaraderie is born out of a shared sense of struggle that you all go through together. There's no way to synthesize that. You've had to fight battles side by side first. And then if you have to go to distributed operations, because people need to be home with their families, fine. You're ready to do it.

Although representing disciplinary identities, the science theme groups are nevertheless somewhat associated with instruments they most often use: the Mineralogy and Geochemistry STG uses the Mini-TES; Soils and Rocks uses the MI and Mössbauer; Geology, the Pancam and RAT; Atmosphere, the Pancam. APXS is directly useful for all of the groups except Atmosphere.

The STGs structured the conversations about the fieldwork among the many team members who do not normally use these instruments in field, collaborate across disciplines in the field, or even do fieldwork. Squyres reminded the team that they needed to work together, and commented, "Looking after everyone's science is the key to keeping morale up during the long, forced march to the hills."[24] One can almost see the group trying to stay together and support each other as they slowly advanced through months. Curiously, the abstraction of all the work as "science," as if it were something being collected in bags, like gold, that everyone can equally value, lends a respectful perspective that might make it easier to be patient and persistent.

Sims remarked that the STGs were also intended to ensure that instrument developers didn't "own a topic," and this freed up the data:

> Squyres created science theme groups as a way to break up the monopoly of instruments controllers and their following. So anyone can be on the geology group. Anyone on the A list [Co-investigators] or the B list [Participating scientists] can join the geology subgroup, and then you can participate in geology subpapers. That organizational rule in part broke up the power structure of the instruments, which was the more traditional approach. I think it's a much wiser way. And all the data belongs to all the team members. There is no private data on the team, from day one.

Des Marais explained that in return, individuals could analyze other people's data—a striking shift from normal academic practice: "It could be different from other scientific efforts in that maybe half or more of the data you really didn't get yourself." The written "rules of the road" regulate this data sharing and publication.

In summary, the science theme groups, the four-sol quartet, and the rules of the road provided a way to shift individual focus from instruments or disciplinary interest to a single, coherent expedition on Mars. Scientists understood that their interests had to necessarily broaden to "figuring out the planet," to working with a team, and to collecting data for future scientists.

Learning to Work Together during Analog Field Experiments

The working relationships among the scientists and the engineers were refined and practiced in ORTs conducted from 2001 to 2003. ORTs brought together tools, organization, processes, and schedules in full cycles of planning, sequencing, and interpreting data. Operations for each sol were performed as planned for MER, although with a smaller number of personnel and a more compressed daily schedule. These tests included training in the "Web Interface for Telescience" tool to define sequences and the "Analyst's Notebook" to access data.[25] Science team members assigned themselves to three STGs—Geology, Mineralogy and Geochemistry, LTP—according to their interests, and the STGs selected their own leader for each day.[26] These tests, which were in many respects actually experiments, were partly driven by the concerns expressed during the PDR about every-sol commanding. Field test training by JPL for scientists had not previously been part of mission preparation.

During the ORTs, the human-centered computing (HCC) team[27] documented a significant lack of coordination among the scientists in their communications with the engineers and in practically carrying out field science through the rover. The tests comprehensively exercised the work system, broadening attention from interface design to issues of communication and workflow efficiency—how plans, programs, and data were created and flowed through the different groups and meetings during a simulated sol. By spring of 2001, the HCC team already knew that the essential work practice question was, "How does a group of scientists do telerobotic science on Mars?"[28]

Scientists' remarks on the first simulated sol in March 2001, with FIDO in the JPL "Mars yard," reveal how, in the serious theater of this mission enactment, software "users" are necessarily recast as collaborators with the engineers in designing tools and processes. Scientists remarked, "This is not a training cruise; this is a shakedown cruise…. They think they are training us; what we're doing is training them in their software…. We are the customer they are trying to find."[29]

The need for naming conventions (and a "naming czar") became salient during ORTs using FIDO during 2002 and training meetings during January 2003. The HCC team concluded, "Scientists don't think like the rover computes or the instruments point." Coordination suffered because the different instruments required different pointing and targeting nomenclature and precision. In particular, pinpointing a target within a feature was necessary for in situ sensing (e.g., placing the IDD directly against a surface for the APXS and MB). But imagery with Pancam was full frame, and remote sensing with Mini-TES required targeting a series of points. Naming is perhaps the most complicated work practice issue encountered during the ORTs, so I discuss it separately in the next section.

The ORTs also revealed many lessons about how data were used and reliably acquired. For example, many people assumed that images provided information that was useful for all of the scientists. But some scientists needed spectroscopy data: "I think in gradations and mixing and color spheres. When I look at an image of a rock, I just see all the same thing." Nevertheless, imagery coupled with a shared feature/target database was essential for working together (figure 6.5): what features can we see on this sol? What targets are we thinking about?

The scientists learned basic logistic lessons, such as doing remote sensing before moving the arm—otherwise, if the movement fails, the program will halt, and data will not be returned on that sol.[30] More generally, the scientists were shown how alternative strategies for specifying Observations, which they defined independently, had different efficiency trade-offs. For example, lumping activities into one formal Observation (e.g., "RemoteSensing PlymouthRock Pilgrim" contained four activities: rover drive; MT; Pancam and front Hazcam) could result in entire Observations being left out of plans when resource limitations and violations occurred. Splitting activities into multiple Observations could sometimes produce less data, but might ensure always receiving something.[31]

In effect, the science team was learning how the rovers could be used, how to develop a joint plan, and how to negotiate the work with the engineers. The 2003 tests taught them how to conduct the SOWG meeting to keep it on schedule and how STGs communicated with the Tactical Activity Planner (TAP). It was recommended that before the SOWG the STGs should coordinate Observations that supported a single scientific purpose and Observations targeting a single feature or specific Activity; this approach helped make the plan's logic and what could be modified more apparent. Similarly, the scientists were asked to document the intent of Pancam and Mini-TES activities so the payload uplink leads (PULs) could properly expand the request into a command sequence.[32]

To facilitate pruning plans during and after the SOWG meeting, the HCC team developed and strongly recommended adoption of a prioritization scheme for Observations and Activities, using a 0–3 scale (must have, important, useful, nice to have). With this scheme, it was less necessary to inspect or discuss the scientists' documented intent. The SOWG plan was initialized by including "must have" and "important" requests and concluded with a "plan ready to go" with respect to resources.[33]

Speaking the Same Language: The Naming Problem

Naming things is an integral part of exploration over large terrains, as people encounter regions and prominent features that they need to reference in maps, scientific descriptions, and plans. Agreements with NASA Headquarters determined how informal names for large features such as craters might be chosen, in particular precluding corporate advertising. But a more fundamental problem was how to name small features that every day the scientists were seeing for the first time and wanting to talk about—rocks, patches of soil, and arbitrary spots among these. Names were required for applying instruments that required precision (e.g., APXS) without burdening other requests (e.g., Pancam) with unnecessary specifications.

This "naming problem" affected the scientific-engineering coordination across all teams and shifts and was identified as a key issue in the development of the science concept of operations during early field tests. Names for rover actions were the basis for planning and coordinating all levels of individual, theme group, and teamwork by both scientists and engineers.[34] (A related, but very different problem involves coding "products" such as the twenty-seven-character file names used for images.[35]) How should the science team specify Observations (e.g., "IDD Stone Mtn piedmont (soil)" in figure 7.3) to communicate among themselves and the engineers? What names would be reminders for evaluating downlinked products with respect to their intentions, that is, why the measurement or picture was requested? Given the expense of each Observation request; the necessity of tracking the work over daily, long-term, and multiyear periods; and the number of people with different disciplinary backgrounds and roles involved, a generally applicable and understandable scheme was essential for avoiding miscommunication and mistakes.[36]

Although it may appear obvious in retrospect for Observation names to combine and qualify objects ("Stone Mtn piedmont") with instrument actions ("MI"), this scheme emerged over three years of experiments and continued to be refined during the mission. For example, a major insight from 2001 ORTs was that most confusion was caused by attempting to pack into an object's name all aspects of the work being done with the object. It was not practical to have a single tag that described all of the "pointing, referencing, and identifying action of the rover, [as well as] situating the work in the remote environment," yet the instrument being applied tended to be assumed in conversation and not documented.[37] A more complex, structured description was required.

The problem of orienting action even when operating FIDO in the JPL Mars yard emerged fairly quickly during the science assessment meeting on simulated sol 2 (March 29, 2001):

> The big part of the work practice is not so much science, but groups are still trying to figure out where they are and to see if they can identify which rock it was that they got held up on. In the long-term [planning] group, someone says, "It would be good if we could just label rocks to orient us." They refer to "round rock." Mike Sims says, "I don't know of a system that does that, so you are asking them to write new stuff [software]." Another scientist says, "But we are trying to go to Mars here." He fills out a lessons learned form.[38]

Such experiences contributed to the development of the SAP software for defining targets by adding the capability to label 2D images using the object-instrument naming scheme.[39]

The naming scheme is an ontology for specifying Observations and Activities, that is, a way of classifying the different dimensions of an operation on Mars using MER (see the sample Activity Report, figure 7.3). Observations describe the instrument and location, and its Activities characterize the method, timing, and focus. Dimensions of the ontology include:

- Instrument:
    - Remote sensing (Pancam, Mini-TES, Hazcam, Navcam)
    - In situ (MI, MB, APXS, RAT)

- Method (e.g., Scratch with the RAT, mini-Mini-TES, Brush, Tau, Drive)
- Survey location (e.g., around, between, including)
- Time (e.g., penultimate, anytime, post-MB)

- Feature (e.g., soil, crater floor)
- Target:
    - Part/whole (e.g., Buffalo eye)
    - Top/bottom/middle (Buffalo top)
    - Other physical descriptors (Buffalo fracture)

In practice, the choice of a physical, action-oriented description was found to be more useful than intentional, purpose-oriented descriptions (related to the "Hypothesis" indicated in the documented Intent; table 7.1).[40] With practice, the team realized that scientists should focus on communicating with the TAP: "Help Payload Uplink Leads understand what you were thinking!" reads the instructional presentation.[41] The TAP combines theme group observations into a plan, working with Science Activity Planner and later Mixed Initiative Activity Plan Generation (MAPGEN)[42] to verify sufficient resources, timing, and priorities. In effect, the scientists are immersed in acting on Mars and need to address others who are not as directly engaged in the inquiry of looking at and manipulating the martian terrain, that is, who don't always share the scientists' experience of virtual presence.

Formally, the naming scheme was characterized in terms of the mission operations as "transfer of science intent through the theme group through uplink," which in effect is a well-known systems design challenge: how are specifications articulated and refined through different roles, representations, and tools? For MER, the rover plan is transformed from initial notations made by the scientists through program sequences generated by the engineers. Both technically and in the collaborative process, the naming scheme helped create one team across science disciplines and engineering. Indeed, an effective—efficient and precise—shared language was an essential foundation for working together on Mars through the rover.

Being a Member of the Science Team

So far, we have considered a kind of top-down perspective on the design of the rover, software, and operations leading to a communal activity. When we ask the scientists to reflect on their experience, they speak very positively about being a member of the team, and they find that the experience has changed them, yet they remain cognizant of the importance of their individual perspectives.

For example, when asked to complete the sentence, "Working with a rover is …," Des Marais replied, "Working with a rover is mandatorily equivalent to working with a team of people. You don't work with the rover without working with dozens of people. Because without several people—a lot of them engineers—you don't go anywhere with a rover, even in a desert here in California. To work with a rover is to work with a large, effective team. That's what I do every day." The forced teamwork of the MER exploration system bridged disciplines, including the science-engineering divide, and thus broadened everyone's participation, sense of purpose, and identity. Des Marais contrasts this practice with the normal approach of field scientists: "You work most with the team when you're out in the field because of logistics. When you come back, you could be working by yourself in the lab. One of the hallmarks of this project is that we're more team-oriented than most, even in the analysis, because we come back at each other, we look for synergies." Des Marais explained that his career success might relate to his being team-oriented, which made him disposed for the collaboration required on MER. He emphasizes, "A lot of research is not team-oriented like missions. Missions are like voyages, oceanography." Furthermore, despite having more than thirty years of mission experience, Des Marais believes that MER's operations involved a radical shift in the scientists' identity as collaborators with a common goal:

> Even though I felt with our previous fieldwork experience that a group is definitely greater than the sum of the parts, that has been demonstrated beyond anybody's wildest expectations to a greater extent with the MER mission than any other I can think of. And so MER for me represents a watershed in appreciating, and acting on, the importance of group-related activities. Each one of us is a classic example of how a bunch of cells getting together can do some pretty amazing things! *(laughs)*

Reflecting on what they've been able to accomplish, he wonders how this might change people's thinking about solving other kinds of problems:

> If we're so darned smart, why can't we put people together and do amazing things? Working on MER certainly will have an effect I haven't been able to fully fathom yet on just how I will do future research projects. In fact, I've noticed that the way in which our activity in Baja, California, has been going forward in the last couple of years is different from what it was before. With the MER activity, I've come to appreciate that there are ways that you work with teams that are more effective. So I think MER's affecting my other research.

Des Marais believes that even the nonscientists who manufactured the rover changed their way of working because they identified with the mission's significance:

> Everything was just done so right, by so many people over such a sustained period of time. It's just amazing. I really believe that there were people sitting in factories, building one of these pieces, who if they had any inkling where this piece was going, stayed an extra hour that day, or they did a little bit more. The cumulative sum of that may be explaining why we're seeing this success. One of Squyres's hallmarks is going around and talking to as many people he could who had a role, and telling them why the mission was important. He went around to every one of the Deep Space Network stations, telling them about the Mars mission, why this was cool. That in the end matters. It's a human process that worked.

Here again we see how the visibility of the rover and how it becomes objectified as the focus of the mission enables people to concretize their role and affects the quality of their work on what might otherwise be viewed as a diffuse enterprise with thousands of people making many "minor" contributions.

But conforming to the domineering group mentality requires giving up some individuality and adjusting to the highly coordinated work, unrelenting schedules, and consensus approach. For example, people sometimes need privacy. The scientists had cubicles where many went soon after meetings. Cabrol commented, "You need that space for being alone, just to be able to think about all the aspects of things, what can happen, and just to be stand back and look at your team and the relationship of things and conceive."

Like Des Marais, Cabrol used the organism analogy: "I really love being part of teams. This has been a great experience, it's very different. You have to immerse yourself and just accept to be one cell of a large organism. And it's not necessarily easy. It does not necessarily come immediately." Cabrol also explained how when she first saw the team roster with the "big names in planetary geology," she was "a little bit scared." She had learned her profession by reading their books; it was an honor to work with people she so highly respected. The experience would change her identity forever: "That was the defining moment, to say we will share something, and this is something nobody can take away from me." Yet the professional level of the group suggested that a very high standard of performance was required. Cabrol continued, "And at the same point I said to myself, 'Oh my god.' Because this reputation is a fact of what you have to be."

Within this group of what Cabrol called "the top game," one also has to stand up to the strong opinions of others:

> You have to have a big ego to do that job. If you don't believe in what you are doing, and think this is really important, you cannot succeed. I'm not saying that you have to kill your neighbor. I wondered, "How is this chemistry going to work? How are those people not going to get at each other's throat in five minutes?" And this goes back to what we were saying about engineers and scientists working together. The distinction totally melts away. Those guys became team members. That's it. I never heard a word louder than the other. The Spirit and Opportunity magic is working.

Carr, one of the well-known textbook authors Cabrol admired, complements Cabrol's perspective by commenting on the confidence shown by "the kids": "A lot of people bemoan the youngsters coming out. I'm impressed by how incredibly knowledgeable these kids are. The kids that we had on MER, you know they were so good! *(laughs)* Intimidating … I mean, my god, they were just…and they took over…. At that age—you know, twenty years old, twenty-two, whatever they were—you're not intimidated, you just act, you don't hesitate, you just do it."

The stark reality is that nobody understands the whole rover or all of the scientific methods and theories. Indeed, people are sometimes astounded to hear what others are thinking or planning. Des Marais reminds us that the rover is so complex that no single person could design, build, or operate it. For example, "No single person could really comprehend whether we were going to succeed in Spirit's landing. When Matt Golombek first talked about the entry descent and landing for Pathfinder, he'd get laughter in the audience: 'You think this is going to work?'"

Rice emphasizes that even among this highly selective group of experts, one person is usually the first to see or interpret something. Thus, individuals remain aware that their thoughts and actions are important, and this ongoing possibility of making a difference reinforces personal sense of knowledge and capability:

> I was the first person to say, "I think we see crossbeds and layered rock" at Eagle crater. That was when we were still on the lander. I remember Arvidson [the SOWG chair] saying, "Rice, what are you talking about?" People thought I was crazy! I spent hours and hours with those images. I would magnify them and blow them up 200, 300, or 400 percent and pore through every little corner of the image, every little pixel, literally, looking for stuff [see figure 7.5/plate 16]. And I said, "We'll see when we get there. I think it's there. I'm standing by what I'm saying. But we'll drive to the outcrop and see if it's there." And it's there! There were a lot of things like that. I was the first person to see the meteorite by the heat shield.

FIGURE 7.5/PLATE 16
Rice working alone, preparing a
presentationin the science assessment
room; Ray Arvidson, SOWG chair,
in the background (February 11, 2004).

Underlying these personal evaluations of how the scientists relate to each other—ranging from awe to intimidation to perseverance—is the ongoing, continuous process of redefining the self, the construction of an identity: am I crazy? Hesitant? Convincing? Making a difference? Grounding these self-appraisals is a shared respect for others' opinions because of their known accomplishments and expertise, which arguably sustained attention in meetings as individuals had the chance to point, explain, and argue. Consequently, despite the multifarious interests and perspectives, being a member of the science team has engendered a camaraderie that Squyres says will last a lifetime: "I love coming back to JPL and just walking these floors in [building] 264. There's this bond on this team. Twenty years from now, thirty years from now, when we run into each other, it's still going to be there. When I go to conferences and see colleagues, you say "Hi," "Hi friend," and shake hands. And then you see somebody who's from MER and it's just like, ah yeah … you know, these are my people!"

Losing the Rovers Means Losing the Team

Signifying what it means to belong to the science team, and how it changed everyone, Rice felt something was lost when he returned home: "I remember when we were out at JPL, it was kind of hard coming back to Arizona because I missed the connection of having everybody in the same room together. We have email, and that's great. But it's not the same as walking down, ten feet away, and just being able to talk to somebody or going to lunch with somebody. I miss the interaction; I think a lot of people do." Indeed, speaking in 2006, Rice felt that the end of the MER missions would be a difficult personal time because the regular routine of caring for the rovers and receiving new data from Mars will end: "When these missions finally are over, when both rovers finally do die, I think it will be awhile before it hits bottom, I mean the withdrawal symptoms. I think initially it's going to be sad. It's over, but you'll be wanting to fire up your computer every day and look at new pictures and it's not going to be there. You're not going to hear anything from these rovers; they're gone."

Silence could occur at any time, though as Cabrol said, they were always particularly concerned about getting to a safe winter haven or "We're going to die." Not just the rovers, but the team's reason for existing would come to an end. In practice, this ending wasn't abrupt or unexpected. The Spirit team had good reason to believe for over a year that the rover was in trouble. On the one hand, the rover had intermittent anomalies (faults and resets) through early 2009, when the team was also struggling to find a route around the northeast of Home Plate. A software update (R3.3) was installed in March and the route shifted to the west. But then another anomaly occurred in early April and the rover became mired in the sand area (eventually named Troy), moving only a few centimeters the first week of May. Although trapped, the rover's instruments remained operational for almost another year. But for about six months starting in late 2009, the team's focus was on improving the rover's angle for winter to a more northerly tilt. Other anomalies and diminishing power dominated the operations, until Spirit finally succumbed to the winter's cold in March 2010.

Imagining the loss of the rovers in 2006, Squyres suggested that a second edition of *Roving Mars* could provide "emotional closure" by describing how they felt:

> The book is not really a science book or an engineering book; it's very much a people story. One thing I hope that you get very clearly out of the book is the passions that drove the people who built these things, who created them and who drive them, and push them, and explore with them everyday. And the deep bond that we have with these two hunks of metal that are up on the surface of Mars. It's going to be a very emotionally powerful moment for a very remarkable extended family of people when these vehicles die. And there will be this process of celebration and mourning that we go through. Like the launch, I think that how it feels at the end of this mission is going to be unexpected. How it feels it going to depend in part on how it ends. If they finally just get coated with dust or Mars finds a way to kill them, it's going to be very different from if we screw up and drive one off a cliff. *(laughs)*

Rice relates his identity of being part of the mission and the sense of loss to the historic accomplishment that they would feel even more afterward: "Eventually, in a period of time, it's going to really start sinking in—what we've done. At this moment [August 2006], we're still ramped up in it so much—the operations aspect, the excitement. It's not really sunk in what these rovers have done, what it's meant, not completely yet.... To me, to be part of the team that discovered lakebeds on Mars—that's a big deal."

It's Still a Job, though You're Working on Mars

What does it feel like when you're in the middle of this complex, historic mission as part of this team? Sims uses the example of the beauty he sees when driving to work every day, where he never stops to walk around: "But in the middle of that beauty, there is: drive the car and get to work. That's what we're up to. We're in the process of working day to day. The structure of a creative team, their nature, is oriented to the work. There's persons doing this, there's persons doing that. You're roughly competing on things. But you get to share overall objectives." He gives the example of a woman who manages the daily plan: "She is a geologist, she loves Mars, but she spends her day, sitting there, at a computer making a plan in the most meticulous details, getting it right, day after day. She does it publicly." In practice, people are preoccupied with mundane details:

> But I suspect given the setup of the team, we're not all that different. Sitting there on a day-to-day basis, we're arguing about power. That's an abstract thing, but you internalize it to something that's called "power"—"Now I've got power for this instrument versus power for that instrument." It's not glamorous or philosophical. It's just, "Is there power for this, power for that? How is this [amount]? Great, and what can I do? And what's a little better?" Later, you can look back and say, "Oh, that was on Mars!"

So your everyday experience and conception of your work and role is not of a historic explorer: "You're doing a job, you're trying to make it work." Only on periodic reflection do you realize, as Rice energetically reminds, "We're on the surface of Mars!"

8

The Scientist Engineers

One evening when I was observing MER operations in Pasadena, I sat with Sims in the engineering control room. It reminded me of NASA's "mission control" at Johnson Space Center in Houston with the desks in rows, like pews, oriented toward walls covered by huge displays (figure 8.1/plate 17). Sims showed me the timeline indicating when communications with Opportunity would occur and what data would arrive. The MER mission manager, Jim Erickson, a veteran of Viking and a handful of other missions and soon to become MER's project manager, came over to investigate. As a stranger in a security-controlled area, I was aware that the human-centered computing team was restricted to observing the scientists and had tried to be unobtrusive. Sims's role was a scientist-liaison within the engineering team, so this one-time incursion seemed justified. Jim remembered meeting me at Ames and greeted me politely, apparently just wanting to establish that everything in the room was in place and familiar.

FIGURE 8.1/PLATE 17
Sitting toward the rear of the surface mission support area (SMSA) at JPL, Sims works as the scientists' representative for the reception of data from Opportunity (science downlink coordinator) (January 29, 2004).

The operations I observed that evening felt like living science fiction—particularly the dramatic moment when I heard the manager command to "Radiate!" thereby transmitting a program to the rover. Yet as I learned a few years later while observing engineers during the Phoenix mission, the calculations and steps involved in actually controlling the machinery on Mars can be a bit dry. Not to diminish the accomplishment, perfect precision, order, and coordination that communicating with the rover requires, but I identify more with the rough-around-the-edges style of the scientists and their inquiry. Still, along those edges one finds engineers who are partnering with the scientists to interpret and enable the work on Mars. So in studying what accounts for the quality of MER's scientific work, I needed to understand the engineers' contributions.

Although the routine looks like clockwork, continuous changes in the Mars terrain, climate, and the rovers challenge the engineers. Without painstaking analysis, experimentation, and invention, the rovers would not have productively survived so long, particularly given the effects of dust, cold, and worn parts. Subject to the many limits of time, power, and troubleshooting at a distance, the engineers' creative problem solving transforms a rover with a broken wheel (or two), a stuck joint heater, dusty solar panels, or IDD frozen forward in deployed position into what is for practical purposes a whole, functional machine. And so from afar, we are presented with a rover that every sol analyzes chemistry, photographs, and trundles along 20 m or more, year after year. The scientists' tidiness will show in their publications; for the engineers, it must be an accomplishment every day.

An Old Class Distinction

For MER, as for all missions, the expertise of scientists and engineers is distinguished with respect to the technology. Scientists determine which instruments to apply in a directed manner and how to interpret the data; their expertise relates to the scientific purpose of the technology. Correspondingly, engineers make, test, package, and control the instruments; their expertise relates to the manufacturing and operation of the technology. As discussed earlier, some members of the science team were involved in the development of MER's instruments (e.g., Pancam, Mini-TES); others are experts in calibrating and interpreting instruments (e.g., Mössbauer). Many of the JPL engineers, particularly those working on the rover and visualization tools, publish scientific papers as well. Generally speaking, the interests and capabilities of MER scientists and engineers are blended, and their work involves related forms of inquiry. For example, Trebi-Ollennu, a rover planner with an engineering PhD, takes pride in thinking like a geologist as he solves rover problems. Trebi-Ollennu, like many JPL engineers, conducts research into the design and operations of planetary exploration technologies, even while he participates in the mission. JPL engineers who worked on Pathfinder's design and operations, such as Matijevic, were involved from the start in creatively blending Squyres's Athena robotic geologist proposal into the MER system.[1]

So to be clear, the discussion here of "scientists" and "engineers" concerns roles imposed by the formal MER organization, in which their domains of responsibility—what they must think about—differ more than how they think. This book focuses on the scientific work from the scientists' perspective, but many of the same topics apply to the engineers, including matters of identity, virtual presence, emotional relation to the rover and the mission's historical nature, and the public and communal experience. My interest here is to examine how the two roles relate and in particular how the engineers contribute to the quality of the scientific work.

Engineering broadly covers the design, development, and testing of the spacecraft and its payload; we are focusing on the last phase, after launch, cruise, and entry/landing—namely, surface operations. Surface operations of a lander or rover occur in two parallel domains: the scientific process of investigating a planetary surface and the engineering process of communicating with and controlling the spacecraft/robotic laboratory. MER scientists and engineers must coordinate their contributions on a strict schedule; cooperate in managing and sharing the key resources (power, memory, bandwidth, time); and collaborate directly at some stages (e.g., the SOWG meeting) and for certain operations (e.g., programming the MER to use a wheel to dig a trench).

Because scientists and engineers are interacting with a common "vessel" and thinking about its operation in different ways with different (though overlapping) criteria for success, it is inevitable that they will develop different identities and approach problems differently. Stories have been told about conflicts between scientists and engineers, particularly on oceanographic expeditions.[2] On ships, a class distinction may develop between those with academic degrees and the mechanics and sailors who run the ship. In part, the ship is the home terrain of the engineers, and the scientists deliberately separate themselves from the difficult, dirty work of keeping the vessel afloat and moving. But the analogy breaks down with MER because JPL engineers as employees of California Institute of Technology are often part of the same "academic class." The MER scientists might indeed not have much collaboration with the Delta rocket engineers who launched the spacecraft, but on the matter of the rover's capabilities and programming, there are many opportunities for collaboration.

Labeling mission participants as "scientist" or "engineer," and the consequential sense of "us" and "them," is in part necessary and in part imaginary, reflecting both important differences in authority and the contingencies of JPL's organization and facilities. Accordingly, developing relationships takes time and is not always easy. Squyres—who came to the project as a university scientist with the authority accorded by a PI-led mission—describes initial shouting matches between himself and JPL project manager, Pete Theisinger, stemming from "baggage we both carried around from past missions" with "stubborn engineers" and "pushy scientists."[3] Throughout the MER mission, the physical turfs and authority of the engineers and scientists are as clearly defined as on a ship, despite the general goodwill and necessity to collaborate.

The Engineers' Role

The separation of floors and rooms, with different key and badge access, formalizes engineering and science into two interleaved but parallel activities that are not jointly planned and not equally prioritized. By controlling the sequencing process, engineering controls the rover. Their calculations of power, data transmission amounts and timing, and so on frame and constrain every aspect of what the scientists do (figure 8.2). The engineers specify when certain instruments can be used and tell the scientists what it will cost. For example, on G36 the engineers declared that any drives must begin after 10:30 a.m. local Mars time, the Mini-TES must not begin until 11:00 a.m. Mars local time, and there is to be no imaging during the rover's UHF transmission to the satellite orbiting Mars; one observation was described as "46 minutes, 41 watt hours, and 2.5 MB [of memory]."

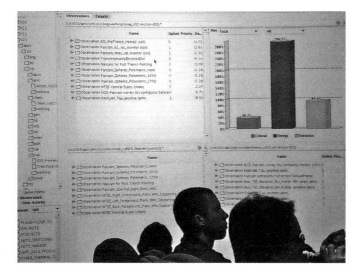

FIGURE 8.2
During the SOWG meeting, a graph of energy and duration summarizes the cumulative cost of candidate observations (February 15, 2004).

The engineers assemble the scientists' requested observations into an actual plan using the SAP and manage the entire uplink and downlink process. Scientist representatives participate in this process, with two positions called the PUL (Payload Uplink Lead) and the matching PDL Payload Downlink Lead). Sims served in both positions during the mission so that he could understand by direct experience how to improve the processes for both the scientists and the engineers.

By analogy, the engineers built the car and are the mechanics and the pilots. The scientists, who contributed to the vehicle's concept and designed or codeveloped instruments, indicate where they want to go, where to stop, and what to do during each stop. The actual plan and route is determined collaboratively. Using software that calculates cumulative power, time, and bandwidth requirements, the engineers can allow only a subset of proposed science activities to be done on each sol (or delayed for later transmission as memory and communication windows allow).

In effect, the details of how observations are accomplished by MER's technology "leak" up to the more abstract level of exploration strategy for determining Mars's ancient climate and geochemical processes. Level "leaking" is an effect in which different system levels interact with each other's organization. For example, if a person eats too much sugar, a "buzz" might be experienced that affects thinking and subsequent actions. By comparison, most designed electromechanical systems are designed to avoid such leaking. For example, today's computers do not require the user to calculate the power, heat, or time required for most everyday uses. By contrast, the MER computer is making actions in the world, in which the environment affects what can be done (e.g., moving in sand usually requires more energy than moving over a hard surface), and its resources are extremely limited. In that respect, using MER is more like computer programming in the 1950s, when use of memory affected how algorithms were implemented. Similarly, it might be like having an electric car that could go only twenty miles on a single charge—you would spend much more time planning your errands and route.

The scientists are continuously reminded of the engineers' concerns and constraints and of their ultimate dependency on the engineers. The engineers' responsibility is balanced by the scientists' control of the planning process—up to and including the SOWG meeting.

The SOWG Room: The Deck of the Ship
The most accessible point of contact between the scientists and the engineers is the SOWG meeting, which representatives of all groups attend. The SOWG room gives an impression of order and propriety, in contrast with the science assessment room. If the science assessment room is a project workroom, SOWG is a room designed for high-level diplomacy, like a meeting of the United Nations. A U-shaped table about 15 × 20 × 15 feet faces the left "front" wall, and another 15-foot table is to the right for the engineers (and one attending mission manager/lead). Large projection screens are all lit by displays (un-

like in the science assessment room, where most screens were not used)—four screens on the left (figure 8.3/plate 18) and two to the right (figure 8.4/plate 19). Four screens show the same small-font SAP windows (the resolution is set too high for observers to read in the far corner). One shows details with a wide-angle image mosaic and simulated lander, with plan locations annotated (e.g., where a photograph will be taken; figure 8.5). The far-right screen shows the familiar computer rendering of Athena on Mars—the same image one finds on the Athena website.[4]

FIGURE 8.3/PLATE 18
The Science Operations Working Group (SOWG) meeting: the engineers are in the back row; the science uplink and downlink leads to left; the SOWG chair is in the center right, his head turning to the right (February 16, 2004).

FIGURE 8.4/PLATE 19
SOWG meeting, view of the corner and right side: the engineers are in the back row; the science uplink and downlink leads sit to the right. The overall plan for the next sol is displayed on the front screen (figures 8.2 and 8.6) (February 16, 2004).

FIGURE 8.5
Photograph of rover IDD (arm) with labeled targets for commanding candidate observations. Photo of display in SOWG meeting (February 18, 2004), used to provide a common reference to observations defined on an adjacent display (see figure 8.6).

The meeting itself can be tedious, as the SOWG chair reviews every Observation and subparts as well as the Activities (figure 8.6 and 7.3). However on M22, the SOWG chair, a senior planetary scientist, was engaged in a playful banter with the SAP operator, an engineer. He said, "What are you doing? The cursor is jumping all over the place. Close that puppy down. Remember that I'm in control." Someone else quipped, "Just keep saying that." Everyone laughed. On M33, the same SOWG chair turned his head partly and called behind him, as if speaking down into the engineering hold of a ship, "Okay, Mr. Op-er-A-tor." By not facing the engineer, he demarked the role boundary between them, and his articulation and tone of voice reinforced their relationship as client (scientist) and server (engineer). As they jostle in this friendly way over who has custody for the rover, their challenges subtly ask, "How much of this guardianship is open for discussion?"

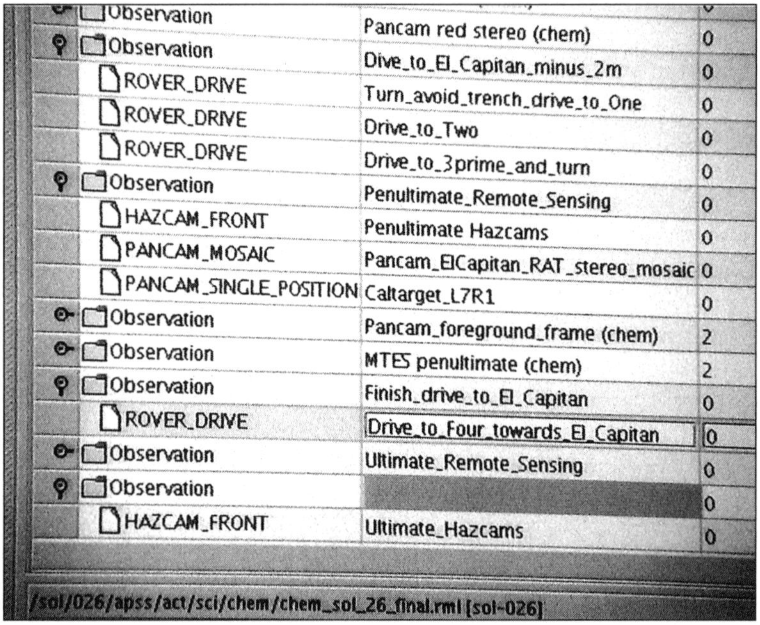

Observation	Pancam red stereo (chem)	0
Observation	Dive_to_El_Capitan_minus_2m	0
ROVER_DRIVE	Turn_avoid_trench_drive_to_One	0
ROVER_DRIVE	Drive_to_Two	0
ROVER_DRIVE	Drive_to_3prime_and_turn	0
Observation	Penultimate_Remote_Sensing	0
HAZCAM_FRONT	Penultimate Hazcams	0
PANCAM_MOSAIC	Pancam_ElCapitan_RAT_stereo_mosaic	0
PANCAM_SINGLE_POSITION	Caltarget_L7R1	0
Observation	Pancam_foreground_frame (chem)	2
Observation	MTES penultimate (chem)	2
Observation	Finish_drive_to_El_Capitan	0
ROVER_DRIVE	Drive_to_Four_towards_El_Capitan	0
Observation	Ultimate_Remote_Sensing	0
Observation		0
HAZCAM_FRONT	Ultimate_Hazcams	0

/sol/026/apss/act/sci/chem/chem_sol_26_final.rml [sol-026]

FIGURE 8.6
Opportunity plan displayed on screen during SOWG meeting, sol 26. An "Observation" is a software categorization for a step in the command sequence, which may involve a combination of driving, photographs, and instrument applications (February 18, 2004).

The SOWG room and the daily meeting during the nominal mission remind me of a ship. The SOWG chair (a scientist) is the captain in a row facing the main screens; the engineering force sits in a row behind him. The big screens are like a ship's windows on the sea (with Mars landscape images showing the rover's location, etc.). The captain is located nearly in the middle of the room; the ship's staff (the engineers) is invisible to him, down under (the back row). The scientists sit around the captain in the large U, like first officers at his side. The session resembles planning where a ship will go the next day.

With its stark black furniture and bright screens, the SOWG room also resembles a war room. The formally defined seating lacks the self-organized, movable feeling of the science assessment room (figure 1.1). A normative protocol hangs heavy in the air, without any sense of affinity—except the obvious grouping of scientists and engineers.

Instead of passing around a microphone or shouting into the air, as in the science assessment meeting, people speak in the SOWG by pressing a button on their individual table-mounted microphone (which turns on a red light at the mount). This public address system is quite good compared to the science assessment room, but it also made speaking up a more formal event.

The seats in the SOWG are of two types: in fixed places around the tables for the official representatives attending and placed in auditorium style to the far-left side and angled into the left rear corner for observers. Scientists in the fixed places on the left are in two rows, the "uplink leads" (responsible for the observations sent to Mars) in front and their matching "downlink leads" (responsible for receiving data from Mars) behind. Large windows behind the Tactical Activity Planner's table on the far right (behind the SOWG chair and engineers) provide a view into the SOWG meeting from a conference room with a huge table.

Overall, the SOWG room appears much more institutional and hierarchical than the science assessment room. And correspondingly, the meetings were more formal. The room is conducive to order and linear decision making, just as the scientists' project room is conducive to informal conversation and divergent brainstorming. Similarly, the engineering command center is the tip of a different hierarchical structure, and it also has been likened to a ship's deck, as related in this striking story about Viking's organization from Gentry Lee, as cited in Ezell and Ezell:

> Gentry Lee also talked of the persons who worked outside the lime-light. Many Viking team members thought of themselves as Earth-bound sailors guiding their ships across the vastness of space. Jim Martin and Tom Young stood in the command center, surrounded by their technical and scientific advisers.... But many others working "in the bowels of the organization" the reporters did not see. Akin to the boiler room crew on a ship, they did all the work necessary to enable the men at the top to pick from several options; they did all the paperwork, computer programming, and system checkouts.[5]

The relation between the MER scientists and engineers is complementary; one knows how to best use the rover for its designed purpose, and the other knows how to accomplish these tasks while ensuring the rover's longevity. The implicit tension is that every requested Observation must be unpacked, scrutinized, and integrated to fit within the rover's capabilities. Yet the matter is resolved as the engineers seize the challenge and take pride in giving the scientists what they want. Cabrol explains:

> The engineers will always be the guardians of the rovers, of their safety, by telling us what is safe to do with them, and what is not. But on the other hand, they always push the rover's limit. What they did was to create and invent ways of going about targets to reach places that was really important to us. In the very beginning, somebody would say to us on the engineering side, "No, we cannot go there; it's dangerous; it's not safe." But on the other hand, another engineer will say, "If we do this this way and that way, we might be able to get there." And they became very creative about what they were able to do.[6]

In effect, the SOWG room and meeting represent a living boundary between the two aspects of the work process. The plan displayed on the SOWG screen exemplifies what social scientists studying cooperative work have called a "boundary object"—it "lives in multiple social worlds and has different identities in each."[7] The relation of the engineers and scientists bears examining in more detail, for it has significantly refined the practice of doing scientific fieldwork with a robotic laboratory on a planetary surface and provides lessons for future missions.

Bridging the Science-Engineering Gap

The complementary roles and routine work with its daily successes have helped the scientists and the engineers to learn about each other and appreciate each other's different roles. The scientists' praise may remind you of the "bonus" interviews that are sometimes packaged with movies, in which the actors rave about the great experience of working with the director. Sims's remarks are typical of the scientists' strong expression of appreciation for the engineering team:

> My experience is that, with very, very few exceptions, everybody I've interacted with both on the science side and the engineering side, are excellent—bright, superb people—with the right motivations, right ideas. And of course there are parochial interests we all have, one way or another. The engineering team at JPL is quite good. Their way of doing business comes out of flights, so it is conservative, and I understand that. They're really good, a really talented crew.

The MER scientists' enthusiasm appears real and is familiar from analog expeditions such as the Haughton-Mars Project. Planetary scientists and engineers are funded on collaborative research projects, such as through NASA's Astrobiology Science and Technology for Exploring Planets (ASTEP) program, to design experimental instruments and use them in analog field settings. Still, the differences in roles, interests, and "ways of doing business" tacitly frame how they think about their interactions. Rather than simply "working together," collaboration is described as "bridging the gap," and the engineer's involvement in the scientific work is not taken for granted, as Yingst described: "I was very fortunate with Pathfinder to work one on one with a lot of the engineers, if you use that term broadly, folks that were doing instrument calibration and that kind of thing. I enjoy those times when I can kind of bridge the gap between science and engineering. How fortunate we are that we have a group of engineers that are interested in the science and want to know what their instruments are doing, why what they did is useful."

Rice says that Squyres did a great job organizing "this vast group of people with diverse interests, the field people and the lab people," and like other scientists, emphasized that MER was different from other missions: "Gray-beards said this is the most interaction they've ever seen between scientists and engineers on any interplanetary project they've ever worked on."

During Apollo, science and engineering worked in different rooms under different security protocols: the scientists in the science back room and the flight controls in the mission control center proper. Squyres set up the organizations differently for MER, having the engineers and scientists present in the pre-SOWG meeting and the SOWG meeting itself (which was led by the scientists). This arrangement promoted communication and a degree of collaboration for difficult decisions. On the Viking mission, and typically during interplanetary missions, team A decides upon and passes plans to team B that has authority to make certain kinds of modifications. MER's organization encouraged the entire team to be responsible for all the science on the rover; consequently, everyone could become involved in engineering decisions with strategic scientific implications.

As an example, Sims described an incident that occurred late in the mission. Spirit's engineering team was ready to turn off the heater on Mini-TES to conserve power. The science team decided instead to temporarily suspend all of Spirit's science activities rather than risk losing its Mini-TES forever. Arguably, this decision might have gone differently or caused contention without MER's working organization and established sharing of scientific and engineering concerns.[8] Rice characterized this interaction as the mission scientists and engineers learning how to think like each other.

Possibly the scientist-engineer relationships have been different on MER than other missions because of the wide flexibility of choices for what to do next afforded by an arm with instruments and mobility. For Viking, the focus was primarily on where and how to dig a sample for the analytic instruments; there were no decisions to make regarding whether to move, the destination, navigation issues, or how to position instruments like the RAT and MI against targets. Rather than working in different places and times, MER scientists were present with members of one of the engineering shifts, who worked with them to manage the daily commanding (again, this organization and schedule was developed and tested successfully in Sojourner operations). Cabrol remembers a turning point early in the nominal mission as the daily uplink/downlink rhythm was getting established:

There was this communication going on that we were trying to establish between the engineering team and the science team. Something happened that was the best thing that ever happened to that group as a whole. We were seeing interesting stuff in science, and while we were there eighteen hours a day, there were hours when we didn't have data; things were slow and we'd have time for presentations. We'd prepare a small presentation and go with a science subject to the engineers when they had nothing to do but wait for data, and explain to them what we'd be seeing, why it was interesting, why we were there. On the other hand, they would come and do the same thing to us.

Thus, in addition to the requirement for collaboration while selecting and sequencing the alternative actions that the rovers made possible, the daily commanding kept the teams working together in the same building, providing opportunities for the scientists and engineers to explain their concerns to each other (compare to Viking's serial schedule in figure 4.1). Besides getting to know each other personally, they were able, to a degree, to assimilate each other's goals and ways of thinking. After several years, the unified team was then better able to handle the more stressful moments that would occur later, such as when Spirit's power dropped to dangerous levels in late 2008.

The issue of moving and staying—an inherent point of contention between the science disciplines—also caused some conflict early on with the engineers who needed to drive to meet mission objectives. Cabrol described the situation with Spirit:

You know engineers—they had put this machine on the surface of Mars, they want to drive with it. At one point, probably on the way from CMS [Columbia Memorial Station, Spirit's landing spot] to Bonneville, the engineers wanted to drive, really, to put some mileage on the rover. They had the rationale, the technology objective to meet, to put 600 meters on the two rovers. We had an interesting scientific objective right there that would require us to stay for a couple of days. Instead of imposing things and saying, "This is the way it's going to be," we took the whole thing, the whole package, up to the engineers' floor, and talked to them: "See, this is why it's important." So we got our two days to do the exploration of that target. And on other hand, when there was a chance that we could pay back that compromise, we did the same thing. This team has been working like that all the time. I cannot say that I have seen any unhappiness.

Cabrol credits the group's shared identity of doing something of historical importance that "goes way beyond any personal interest or even group interest. Just being part of it, and making it work is the only thing that people were interested in."

MER's formal mission objectives, mentioned by Cabrol, do provide a degree of objectivity that on its face might suggest that engineers could act alone. The "full success" criteria, sometimes labeled "metrics" on presentation slides, were:

- Operate both MERs for ninety sols (or two MERs for thirty sols simultaneously)
- Drive 600 meters with at least one rover
- Use each of the Athena payload instruments at least once on both MERs
- Create 360-degree color and stereo panoramas

- For each MER, take one image of a rock to which the RAT had been applied, with complementary analysis by at least one instrument
- Investigate eight locations

During the nominal mission, a scorecard showing to what degree the "full success" criteria were met was typically displayed and reviewed during the morning science context meeting. At the root of the conflict mentioned by Cabrol, MER's mission objectives say nothing about the scientific understanding of Mars. For example, a mission objective might have stated, "Determine whether Gusev crater was once a lakebed."[9] Similarly, "complementary analysis" refers to applying another instrument to the same location, rather than that the data must be scientifically useful.

Because MER was defined as a scientific mission (in contrast with, for example, Deep-Space-1, with its dozen advanced, high-technology experiments[10]), the engineers had to delay meeting the "mission success criteria" so that they could satisfy the scientists' objectives. This necessary compromise, particularly regarding the drive metric, which wasn't satisfied for Spirit until sol 89 after it left Bonneville (table 2.2), led to an ongoing communication of concerns and with time the working understanding that Cabrol described. With patience, diligence, and some luck, both the scientific and engineering objectives were accomplished.

As the rover aged, new challenges arose, requiring the teams to reestablish priorities and ways of working. But Cabrol said in the summer of 2006, "After something like 920 sols, living with the rover, the science team has a pretty good idea what is dangerous or not to do." And the team developed a humorous pleasure in working with the rover's changing capabilities: "And you know what we said, the day we lost the mobility of the rover, we didn't say we lost a wheel, we said we actually gained an autonomous trench-maker! So this is the spirit of the team, literally. Take what you have, not what you wish you had, and do the best you can with it."

The Engineers' Concern for Science

The scientists make special mention of engineers who show unusual concern for the scientific work. Squyres credits Gentry Lee, the Viking veteran who made important contributions during the PDR, for helping him formulate his principal investigator responsibility as science systems engineer:

> Gentry is one of the best systems engineers who's ever lived. But he is also very unusual in that he has a truly deep appreciation for the science. He really cares about the science being important. He understands what's important to science well enough that he can look at the design, and say, "This, this, and this are under control. These things are out of control, but they're going to get fixed by smart engineers, and you don't need to worry about them. But this item, Steve, and this item, you better devote attention to, because they're going to come back and bite you in the ass if you don't."

So for example, when I started on MER I had no experience with actually developing and building flight hardware. I understood some of the engineering, but I hadn't actually taken a piece of hardware through the whole development cycle from initial concept all the way through to launch. So I had never been through thermovac [thermal vacuum tests] before. Gentry came to me and said, "Look, Squyres, something you better understand, you've got an enormous amount of testing that you're going to have to do in thermovac, and right now the schedule does not have enough time for that. You better scream bloody murder about getting that time in thermovac, because if you don't, essential calibrations that you need aren't going to get done." And he saw that way before I did.

Squyres followed Gentry Lee's advice and subsequently preoccupied himself with systems engineering with an eye toward the scientific implications. Squyres mentions other engineers, but calls special attention to Randy Lindemann, mechanical lead for the rover: "What made Randy special was his passion for science. It wasn't just that he wanted to build a rover that could do good science on Mars. A lot of engineers felt way, I was coming to realize…. The thing that set Randy apart was that he was genuinely curious about Mars itself."[11]

Trebi-Ollennu was part of the original Athena team with Squyres and participated in the MER proposal. He describes how he's worked with the scientists from the start to make the rover useful for field science: "When Steve [Squyres] and Ray [Arvidson] were developing the proposal for MER, we were doing field tests and training. We came up with a training program for how to train scientists to do 'geology' using a rover. So I'm very close to those guys. Most of the Athena science team, I know them for years before we landed on Mars." How has working on MER changed his identity as an engineer? "First of all, as an engineer it's strengthened my skills and my confidence. On the other hand, it's given me better decision-making skills, working in a multidisciplinary team. I'm not a scientist, but these days when I communicate with a scientist, we're on the same level. I understand what they're trying to do, and I can effectively communicate the capabilities that we have to meet their needs." Paralleling the interactional pattern described by Cabrol, he explains how scientists and engineers worked together:

Usually, when the scientists say, "Okay, we want to get to that rock," when I was less experienced I'd say, "Okay, let's drive to that rock." But these days, I'll ask the scientists, "What exactly are you looking for when we get to the rock? Do you want to acquire a Möss[bauer] spectra or do you want to acquire a microscopic image to look at the texture of the rock? Because then I have to park the robot differently, so you don't have shadowing or you don't have half shadow, half light. You want to have consistent lighting. On what day and what time of day do you want to make the observation? Because that makes a big difference." The other thing I'll ask the scientists is that if we're going to stay here for this number of days, "What type of data are you going to acquire? And we might have communication problems. So let's check to make sure that over these following days we'll have enough bandwidth for us to get all of the data that you want down."

By sharing information about his own decision making, Trebi-Ollennu pulls out in conversation what variations will make a difference to the scientists. Like an empathetic travel agent, he helps them to articulate their motives and constraints while revealing available options, so that together they develop a path and plan for the rover. Although Trebi-Ollennu began with no particular interest in space science or geology, he found himself getting more interested in the geology of Mars:

> Yes, I am interested *(laughs)*—surprisingly. You know, because the scientists are interested in layers, anytime I see a layered rock, I get fascinated. "We've not seen these layers before, why don't we take more pictures of this layer?" And you tend to almost do the science. You have to remind yourself that you're not a scientist. And we've been able to point them to unique rocks, saying, "This looks very different, let's take a look at it." And actually one of the best rocks that we looked at with Spirit, which was called End-of-the-Rainbow, was discovered by one of the rover planners.

When this event was recorded on the JPL public website,[12] the engineers' contribution was invisible, just as the particular scientists involved are not mentioned:

> Spirit Reaches the "Columbia Hills"!
>
> On sol 156, Spirit roved 42 meters (138 feet) closer to a vantage point where it could observe the hill outcrops. Some of the images that Spirit sent back revealed a small and unusual *rock that piqued scientists' interest* and was informally named "End-of-Rainbow."

Consider how strange it would have appeared, given the style of these somewhat dramatized reports, to report the truth, that "the unusual rock piqued the *engineers'* interest."

Matijevic tells the same story of how this rock was found, personifying the joint science-engineering team as "the little rover that could":

> Here's poor Spirit, you know, spending all its time trekking across the terrain, *(laughs)* across Gusev, looking for something that wasn't basalt! *(laughs)* And thinking, "Well, maybe there's something that's up this hill." And slogging up this hill, and going through the winter season, that's always kind of hard because of Gusev being relatively far from the equator, then the conjunction inhibiting communication with Earth, and suddenly there's slipping and sliding all over the place, and all the rest of it…. And you know, somebody just looking at the images and saying, "Hey, this looks kind of interesting!" *(laughs)* It doesn't matter that it was an engineer who said that …And suddenly, there was this discovery of outcrop rock that we had been looking for at this particular level.

Trebi-Ollennu gives another useful example about how the engineers' experience enables them over time to form a real partnership with the scientists:

> After working with John Grotzinger for a while, I got to understand what he was looking for, the way he was trying to acquire the microscopic image of the festoons [in which water forms in the layering of rocks], so he could reconstruct them and compare them to rocks on Earth. So I was able to explain to him what kind of "approach normal" [having the boresight of the instrument at a vector of 90 degrees to the layers of interest] we could use to get the information he was looking for. So we could say, "Okay, these are the parameters that we can tweak to meet your needs." This was published in *Science*[13] as one of those fascinating finds on Meridiani.

Matijevic, having a more encompassing role as engineering team chief for MER Operations, is less involved in day-to-day targeting decisions but nevertheless has gained a new perspective about the instruments and the logic of comparative geology:

> I clearly learned a lot. I'm not a geologist and certainly didn't start off knowing about the kinds of things you'd want to, that you *can* do with some of these instruments, much less what the integrated effect of these instruments will tell you. I also learned about the limitations as well, the things that you can't really do with this generation of instrumentation at Mars that you'd really want to do, in order to answer some of the questions. And I understand the logic these people go through. We make assumptions that the geology at Mars is like the geology that we have at Earth—so you look at something and you see the elemental content of a rock, and pull out your handy-dandy list of Earth-like rocks and say, "Voilà! This is olivine!"

Strikingly, Matijevic usually refers to "the vehicle," not the rover, the robot, or MER. For example, he said, "With a vehicle like this, the observations that you can make and the investigations you can perform each day are a function of what you can reasonably do with the energy and the mechanisms that are available to you." So not all of the engineers have the same perspective on the technology. Matijevic started as a mathematician; for Trebi-Ollennu, who has an aerospace engineering doctorate specializing in robotics, the robotic (programmable) aspect of the vehicle was salient. The conceptual difference is that a person drives a vehicle and a robot drives itself. This distinction illustrates that the variety of specialties and interests among the MER engineers is at least as great as among the scientists.

Driving a Vehicle in the Test Bed

The scientists tell us that their orientation of "being there" on Mars is in terms of "What would I do?" Correspondingly, the engineers' orientation is "How do I get the rover to do what the scientists want to do?" Besides the Science Activity Planner (figure 8.7/plate 20) and related computer tools for formalizing "observations" into plans and programs, the engineers strongly rely on physical test beds, in which they take a duplicate rover through its paces on simulated Mars terrain or in simulated thermal conditions. Matijevic explains that the choice of tools is also a matter of engineering background: "Many of our engineers have computer science backgrounds. To them, the visualizations that are possible within simulations are fine. They don't necessarily have to be at the place to see these things. For other engineers seeing the vehicle, seeing it do certain things, is the way they create mental models for what's possible to perform." Trebi-Ollennu suggests that this approach develops from experience: "Going to the test bed with my colleagues and coming up with a workaround for a problem is I think one of the things that they didn't teach me in school."[14]

Matijevic explained that there were two test vehicles. The Surface Systems Test Bed (SSTB) vehicle was used for mobility, navigation, and tests such as one-wheel dragging. The other test vehicle, SSTB Light, with electronics removed and controlled by a buddy box (operating by a tether to a remote control unit), is roughly one-third the weight (50 kg); it is used in the sandbox and for combinations of tilt and environment testing. Other rovers similar to MER were used for the ORTs, which occurred in a variety of environments.

Learning how to interpret the status data received remotely so that one can properly grasp the reality of the rover's circumstances and understand how it behaves in different conditions was important for everyone before attempting to control the rover on Mars:

> One of the really beneficial things that you have in a rover mission is that you literally can build your rover, you can put instruments on, you can send it places—here, on Earth— and let it feed back information, so you get a sense of what you can see and what you can't see. We've done everything from lab tests, to Mars Yard tests at JPL, to field tests in the Mojave—all really meant to give people a notion of what that model is like, to realize what you're doing. We sent just about anybody who was available in those days, scientists and engineers, to these sites to see what they saw with the pictures that were taken, versus what was actually there. That was an important mental model to develop, to do that correlation.

FIGURE 8.7/PLATE 20
Using the SAP tool, scientists define targets and instrument operations in the Meridiani science assessment room to plan RAT operations and photographs in Eagle crater (February 19, 2004).

Other early simulations showed how the instruments work together, how controlling them fit into design of mission operations, and the logistic engineering steps in which the Observations were embedded (e.g., turning on and off the IDD heater).[15]

When asked to complete the sentence "Working with a rover …," Matijevic focuses on this sometimes jarring experience of imagination confronting reality: "Oh, it's just humbling. I don't care what you think you believe you understand about moving a vehicle, it's never like that. As soon as you're on a different terrain, it's a different regime. I think I've learned as much from driving a vehicle into the sandbox, and learning about what's possible to do, what can't be done. *(laughs)* It's not like driving your car." Matijevic says that in setting up another team for another mission, considering the requirements of facilities, computer tools, staff, and so on, he'd emphasize the importance of the vehicle test bed:

> You can't downgrade the contribution that comes from having an actual, full-size model, putting it in different situations, and just watching that vehicle perform. Even if you have to buddy-box it. You literally see the vehicle's capabilities, as well as the challenges that it has on those circumstances. That's just an experience you don't have a way of replicating. We use that all the time. Most of our operation is in one fashion or another traceable to a particular test that we performed in a sandbox.

For example, Opportunity's operations on the steep face of Victoria crater relate to tests preparing for entry into Endurance crater. The team "spent several weeks prototyping operations in various levels of tilt and various kinds of terrain conditions," which provided experience and methods for later operations on slopes.

The Engineers' Problem-Solving Expertise

Although some engineers read geology in their spare time and others speak as eloquently about exploration as any scientist, their most proud stories are about solving practical problems in operating the rovers, not "figuring out the planet." But to be clear, the engineers' orientation has a "both-and" aspect—their satisfaction derives from both keeping MER healthy and carrying out the scientists' plans, which includes suggesting opportunities for observations and finding new ways to do them. The identity of the rover engineers and planners is framed by the scientific enterprise; they are not merely operating a ship.

For example, Trebi-Ollennu describes the engineers' pleasure in using their skills to accomplish what scientists thought might be impossible:

> We went to Hillary—it's on a very big, high slope, about 28 degrees—to do robotic arm activities when we were on top of Husband Hill. It was published in *Aviation Week* [figure 8.8/plate 21 and figure 9.4]. It was a very fascinating period. The rovers were not designed and trained to do in situ work on that kind of slope. To the scientists, it was almost unimaginable that we could do that. But because of our skills and tenacity, we were able to push the rover and push our skills that far—by not taking unnecessary risks, but very systematic analysis to demonstrate that we can do that.

FIGURE 8.8/PLATE 21
Spirit at Hillary Outcrop on Husband Hill, September 2005. *Image credit:*
M. Di Lorenzo et al., Mars Exploration Rover Mission, Cornell University,
JPL, NASA. *Source:* NASA Images, "Attacking Mars," http://www.
nasaimages.org/luna/servlet/detail/NVA2~4~4~6297~106823:Attacking-
Mars (accessed July 23, 2009).

Through clever, systematic problem solving, the engineers were also able to resolve a problem with Opportunity's instrument arm:

> On Opportunity, we have a robotic arm anomaly. One of the windings of the DC motor for the azimuth joint was broken (opened). When we tried to unstow it, it wouldn't unstow. We were able to detect the broken wire by good analysis—we were able to find that a wire was broken one hundred million miles away. And I led that team. Without the robotic arm, we lose two-thirds of our science payload. So we had to recover that for science. Steve gave us almost two months, just to do that. Can you imagine just trying to find a broken wire in your car one hundred million miles away without any images, just telemetry?

Trebi-Ollennu describes the detective work required to fix the rover. At this point, he was investigating an unknown system, developing a structural model that explains its behavior. Using diagnostic troubleshooting methods, he formulated experiments that tested different explanations of how the system was configured and how the parts were interacting:

> The first thing that we did was send our regular commands to the system and get the telemetry back, and look at the voltage and look at the current. After a while we could see that we were applying a lot of voltage and we weren't getting a lot of current out of the motor. So we realized we had added a resistance. Obviously, the resistance must be getting higher. So the question is, "What has changed the resistance? Is it debris? Is it the lube, the dry lube, or the wet lube, lubrication that we have in the actuator? Is it something that is impeding the rotor?" So we went through and actually came up with several hypotheses. For each we asked ourselves, "What data do you need to verify or eliminate it?" And we went through a systematic search to eliminate all the hypotheses until we settled on one.

So the reasoning methods of robotics engineering and geology are similar, using hypothesis testing in a field setting: "That's why I think Steve and the other geologists are able to get on with robotics, because robotics is an experimental science. And that's geology as well. It's an iterative method: you get data and then you iterate, you keep going until you pinpoint your solution. We have similar skills." Trebi-Ollennu continues the story of how the arm's mobility was recovered:

> We realized that we have a broken coil in the windings, because the motor rotor resistance appears to have doubled. What do we do? We were able to change our voltage limits to increase our current draw, and it started working again. So, since sol 654 we've had this broken winding—we've been able to use the robotic arm with a broken winding. I can't believe it. We have to pinch ourselves to see that we can use it, that that's what we do as rover system engineers.

Such fascination with the combination of observed facts, the explanation, and the new method may be replayed over and over in an engineer's mind. The incredible causal story requires repeating, with each retelling reigniting the pleasure of the logical closure and underscoring the lesson in memory: "We have an opening in one of the windings. And the reason that we have the opening is that when we landed with Opportunity we had a stuck on heater on the joint. It was seen going from +70 to –70°C every day, a huge thermo-cycle that it wasn't qualified for. To me, what was fascinating was that we could diagnose the fault, come up with a workaround, and get it work!" To have solved a problem that might have crippled the mission, in a manner that worked for years, for a mission slated to last only ninety sols, is indeed impressive. Trebi-Ollennu is telling us how MER succeeded, how the mission has gone on so long, and especially how working with a rover is not just a matter of sending up commands that satisfy the scientists' needs, but learning how to work with the rover you have—one that is changing as the months go by:

> Another example was with the Rock Abrasion Tool on Opportunity. It has three actuators. They affect abrasion in a planetary arrangement: the grind actuator rotates the grind bits about the bits' center; the revolve actuator orbits the grind bits about a common center (forming a big circle with several concentric circles inside); and the Z-axis actuator controls the penetration of the abrasion area. All of a sudden the encoder of the grind actuator failed. The encoder tells us the position of the actuator, so we had to find a workaround. Because we didn't have position knowledge, we couldn't use our high-level commands, we had to use our low-level commands. And in our case, for the grind encoder to fail is the best that could happen to us, because we can apply a voltage and basically just drill [abrade in open-loop fashion]. But unfortunately, a few months later, the revolve encoder also failed. That posed a completely different challenge—the revolve actuator orbit motion requires closed-loop control [because it varies the orbiting speed based on the torque at the grind bit]. So now, you're one hundred million miles away, and you're trying to perform coordinated control of three actuators without the requisite feedback, so we resorted to using time. We turn on the grind actuator and run the revolve actuator at a fixed orbit speed and step-down into the surface with the Z-axis at fixed time steps. We effectively bypassed all the fault protection for the RAT. This was a very bold step since the RAT was the only tool on the robotic arm (IDD) that could literally break the IDD. That's what we've been doing since sol 1045.

Trebi-Ollennu's stories about Husband Hill and Opportunity's arm and the RAT—with his precise memory of specific sol numbers and the procedures they tried—focus on the engineering details of delivering the rover's designed high-level behaviors (e.g., driving, drilling) by improvising low-level control programs. For sure, he is aware of the larger scientific context (illustrated by his remark about *Aviation Week*) and values his own work according to its contributions to the science. Although his primary excitement is not from discovering that features of Mars were formed by liquid water, he has assimilated these interests, and his identity has blended with the scientific purposes. Asked to describe his disciplinary specialty in light of his experience on MER, he replies that he's "still a robotics engineer, but I'm a robotics engineer that does in situ exploration, planetary surface exploration." So like the scientists, he's an explorer? "In a funny sort of way, I think of myself as an explorer, but doing in situ work using robotics. Because that is where my emphasis is, how do you get robots to do in situ activities to serve science?"

Trebi-Ollennu's engineering interest began with an awareness of aerospace automation around age seven:

When I was growing up, I lived very close to the International Airport in Accra, Ghana. The flights came over, and I was very interested in being a pilot. I knew the names of all the plane models. I was very interested in doing something in aerospace, mostly of trying to fly an aircraft or be a flight engineer. I used to write to British Aerospace and got all the brochures about jet fighters. And that's how my interest developed. I went to Queen Mary College to study avionics, because I was more interested in the autopilot system. I was fascinated about how to get humans out of the cockpit and get computers to fly the thing.

So what is it like to work with a rover, getting "computers to fly the thing," for so many years? Why have engineers with a research orientation like Trebi-Ollennu stayed involved in operations?

There's an allure to this job that you can never explain. I think surprisingly, most of my colleagues find being a rover planner a very prestigious thing to do. Unfortunately, I don't think I do. What I found about myself that surprised me is that I enjoy working on anomalies. As one of my colleagues said, "It's just like *CSI* [*Crime Scene Investigation*, a television show], trying to investigate a few million miles away a broken wire." I think I enjoy that more than driving a rover. It surprises me! Trying to find a solution and work-around for problems is what I really enjoy. I just found that out. I thought that what I was really interested in was getting a robotic arm to pick up the science and do all these things. But with time, I think I'm beginning to see, "Oh, I just like to solve problems." And no matter what's the nature of the problem—if it's a science problem, the scientists just want to get this data; I just want to solve that problem for them. I think that's what it comes down to. But I'm sure if it were just a circumscribed routine—just driving 5 meters there, 2 meters there—I think I would have left the project by now.

Well, then, as the rover is getting older and more parts fail, he's surely coming into his prime? "Yes. *(laughs)* I'm just looking for a job where I can solve problems." Unlike some of the scientists, Trebi-Ollennu didn't struggle to find a role—his tasks were well defined because they center on the rover's fundamental behaviors, some combination of which is part of every day's work. Yet like the scientists (and perhaps everyone involved in the mission), the longer-term experience of being part of the mission, of participating in an engaged way and consciously not moving on to something else, is revealing for each individual their personal qualities and values—what they truly like to do and what they are good at.

Matijevic provides a similar perspective; his job is fun because "There's no lack of problems that we face from one day to the next here. Being able to be in a position where I can contribute something to their solution is quite a sense of accomplishment."

So what about engineering teamwork—suppose we were advising a group in another country who were just starting out and putting together a new team. What should we be sure to emphasize in the design of engineering operations? Trebi-Ollennu would emphasize "the diversity of the team—not in terms of ethnicity, but in terms of skills, in terms of personalities." In particular, one should not primarily choose people for engineering roles because of their background, but according to their skills:

One of the enriching experiences on this mission is that I would have naturally thought that everybody who drives the rover and controls the robotic arm should be a roboticist. But the surprising thing is that among the rover planners, we have a mixture of roboticists who are PhDs, and we have people who developed the flight software and the ground tools that we're using. Having this diversity has really been good. Because when you're using the tool and you have a problem with the tool, they can fix it, and they can show you what you're missing. And they can train you how to use it. Similarly, after testing with the robotic arm and the motor system, I'm able to impart that knowledge to others. So the skill diversity is really surprising and it works out really well—it is very important. Having people with a science background on the rover team I think will be even better. You shouldn't just say because somebody has a certain background they might not fit; they bring a different perspective and it enriches the team very much.

Matijevic explains how the engineers' diversity relates to different aspects of problem solving during operations:

We can't get by without the specialists, and we can't get by without the generalists. These are team participations—a combination of the people who are going to suggest what kind of problem has to be solved, people who are going to tell us how we can fix something. Others are actually going to test out the fix on our test bed, and others are going to build the sequence products and the like that are going to go up to Mars, to actually *implement* the fix. That's a *group* effort. And you really needed all those disciplines to make that possible.

Trebi-Ollennu comments that in addition to expertise, the team benefits from complementary personalities for relating risks and opportunities:

I think pairing the right people is always very important. In our team, we always have two rover planners on shifts. Having the two people with different temperaments makes the sequences more robust: somebody who likes to take a little more risk and then somebody who's more cautious. They always come to a better compromise. There are certain teams that don't work well together because they're hyper or all type-A personality, and you tend to have more errors.

So how would he summarize what it's like to work with a rover? After a three-second pause, he laughs and says, "Working with a rover is a lot of fun." Told that he agrees with Squyres, he replies, "Oh, we do? *(laugh)* It's a joy, actually."

9

It's after 7:00 p.m. in the Haughton River basin on Devon Island, on July 22, 1998, our expedition's last night in the Canadian Arctic. Everyone is gathered around a table piled high with bread, condiments, and Tang, in the heavy, white canvas "mess" tent we use for cooking and eating. A brilliant yellow-orange light glows behind the west wall, partly shaded by a cockeyed Mars Pathfinder poster but still brightly illuminating and adding some warmth to the only occupied space within 100 miles. Mike Long, a writer from *National Geographic*, has assembled the group for a poetry reading of the doggerels he has composed for each of us:

> **Long:** We'll start with Dr. Rice—
> Rough and ready, macho Jim,
> Explores the crater, rim to rim,
> Examines rocks that teeter totter,
> Bench presses ATVs in water.
> Has his eyes upon the stars,
> Will he really go to Mars?

After ten of these and much applause, there follows a silence, after which Jim Rice starts the group on a quietly spoken, more personal sharing of experiences:

Rice: Mike, what's your most vivid memory of Haughton?

Long: When I spun off that ATV, that's what first comes to mind. That was my first day out, following Pascal up the hill [to Sapphire Lake].

Rice: And what's the next thing that comes to mind?

Long: I think sitting in middle of the Haughton River [on an ATV], watching John [Schutt, camp manager] perform his bucking bronco act … *(laughs)* on the water.

Pascal Lee (Haughton–Mars Project expedition leader): That was on our traverse south.

Rice: Yeah, that was the time I came…. What about you, Peter? What's your most vivid memory?

Peter Essick (*National Geographic* photographer): I liked when I went out late at evening…. Like at midnight, just cruisin' around. You feel like the whole crater is your own. Has sort of a whole different feeling, out there by yourself…. I liked going on those traverses, that was always fun, and … hmm, some of the science things, when we were out there and it was snowin', really cold, we were all huddled up, with hoods, down there by the rock pile….

Rice: I guess if I had to pick my favorite area out here, I'd say it's Sapphire Lake and Aqua Vallis, the big alluvial fans. There's a lot of stuff all compacted into one small location, lakes, valleys….

Rick Alena (NASA Ames computer engineer): For me, it was discovering Polygon Lake with Charlie [Cockell, biologist] … deciding to see what was in that next range….

Rice: That's fun, to see what's just over that hill or around that corner.

Alena: And then looking in and seeing there's water….

Rice: What's your favorite place, Pascal? If you had to pick one.

Lee: My favorite place? Ah … I'd say I have a couple, one is the site that's right above camp there, on the breccia—you can see the whole valley. Another one is at the top of Tripod Hill.

(All): Yeah!

Gary Martin (NASA flight surgeon): That's my most favorite one. Without a doubt, it's amazing up there.

Alena: It's an unusual view … that you don't really expect to see, the rocks, shapes, and things….

Lee: Spectacular.

Rice: John, what's your favorite place out here?

Schutt: I hadn't really considered that; in truth it's all nice. I guess my sleeping bag….

Lee: What about you, Mike?

Long: Tripod Hill's my favorite place, especially to the west. The river, the configuration—the breccia and the hills—the landforms…. It's a spirit place.

Poetry, favorite places, talking about being alone, the romance of the breccia—during that ten-day Mars analog expedition with a dozen men, I never heard a single conversation about sports, cars, or women. Sitting with the MER scientists during off-hours at JPL, the focus was also on the immediacy of the experience, with an almost reverent feeling of presence, our talk centering on "what's impressing me" and "what I'm doing here, now, in this place" and how it all relates to our shared vision—our "eyes upon the stars." So in interviewing the scientists, I wanted to hear how MER related to their personal lives, including what else they were doing and how they experienced the mission aside from being an explorer or finding a niche. What other creative impulses drive them? How do aesthetics and feelings relate to the scientific work or, indeed, the idea of going to Mars? How does the mission affect them emotionally?

Private and Public Aspects of Scientific Identity

People develop an identity by interacting with other people as they participate productively in activities of mutual interest. Developing a sense of self is therefore inherently social as well as personal. Because the scientists have different disciplinary backgrounds, organizational affiliations, and funding sources, they are always juggling and blending social perspectives on "who I am" and an ongoing sense of "what I am doing now" as they realize their interests (both in deed and in understanding). Even when scientists, inventors, or artists are working alone on personal matters, they are often tacitly relating to what they will show others or how their creative products will be valued and how they will be received.

The identity of the MER scientist naturally focuses on what it means to be a member of the science team. As preceding chapters have described, the professional identities of the MER scientists are adapted for the needs of the mission and teamwork, yielding an experience of participating in the science team (communal identity), an experience of being on Mars (explorer identity), and an experience of scientific discovery and theoretical advance (intellectual identity). All of these perspectives develop more or less simultaneously in interwoven threads within activities mediated by the robotic laboratory technology.

At the same time, the scientists have activities and interests peripheral to their roles on the mission. Some of these concerns are *private*, based on a sense of a career and intuitive interests and capabilities; simply put, these are "internal" motivations. Other concerns are *public*, fostered by the global awareness of the scientists' work and contacts from people outside the mission; these are "external" motivations. As for all aspects of identity, despite the apparent origin (internal or external), the emerging conception of the self is social, relating the character of the individual and the group: private concerns are given meaning by reference to social values: "How am I contributing to my community?" In turn, public interest shapes how the scientists value mission events by redirecting their attention to images, findings, or milestones and reinforcing their historic context.

This chapter illustrates a range of private and public aspects of the scientists' identity that are apart from operating the rover, reflecting their broader sense of individuality and being a member of society: personal projects outside the mission, personal experience as public figures, aesthetic motivations in creating images of Mars, and emotional experiences. In this analysis, stepping outside mission operations and looking back in, we also gain perspective on how people not involved with MER may be drawn into the adventure by witnessing the scientists' excitement and, particularly, the bizarrely familiar images of an alien planet.

Personal Projects

Each of the interviewed scientists sustains a personal project or interest that they view as a seed for the future. Generally speaking, a researcher's professional career is a kind of *enterprise* with multiple projects and interests. Multiple facets of personal, professional work are multithreaded, interwoven, in funded, institutional projects in different ways over time. For the individual, the long-term experience is like juggling balls in the air—if an interest is allowed to remain undeveloped for too long, a project is left unfinished or a thread of interest not developed, at some point the person must acknowledge that perhaps he or she will never come back to that activity, and thus has lost that identity.

For example, Cabrol mentions that while participating on MER she is reminded of her personal interests: "While the teamwork is a very important part of the magic of Spirit and Opportunity, and I want to continue to be part of it, I will definitely continue to develop my own research and try my own concepts, and test my own stuff. This is what I feel as a person … testing those ideas in the Earth system context, and how these ideas can explain our own origin and what's going on in other planets. I am hoping that I can contribute there, too."[1] In considering the Earth and planetary science more generally, Cabrol places her work on MER and other missions within a broader inquiry that she conceives as her personal intellectual project.

Given the consuming nature of MER Sims described, it's somewhat surprising that he works on other things, too: "I spend lots of long weekend nights doing modeling of physics. Software modeling of physics. I'm interested in the process of model discovery and refinement."

What motivates this personal, private work? Is it important because everyone needs something to call his or her own? Or because scientists are brought up in an academic culture that requires developing unique identities, as signified by PhD dissertations? Because MER brings people around to doing science in a different way, through what might be called a "group mind," is there a need to balance the demands of this work by cultivating a private inquiry (in Cabrol's terms, creating a space, an activity, for being alone)?

Sims partly answers this question for himself: "It's probably because those personal projects are part of my life, I'm much less concerned about actually proving in MER that I'm doing things that people would consider research." Thus, Sims expresses the norm of what it means to be a computer science researcher and believes that his personal work establishes his productivity and, indeed, his inherent claim to being a researcher. Like the other scientists, his personal project reflects a long record of past interest and investigation.

Cabrol has led field expeditions during MER: "I disappeared from the MER team for between four and six weeks." Asked how she felt about having two projects, she said, "It's like having two kids, which one of them do you prefer? *(laughs)* There is no answer to that; you love them both. Some need attention at some point, others at other points."

In explaining how people become engaged in creative projects, Tim Hetherington, a photojournalist, said, "It's got to come from yourself … that's the most honest place.… If I started saying it came out of a desire to change the world, that's very suspect. Can't it come out of a place of personal curiosity? A desire to locate myself in the world and also to have some utility?"[2] This statement nicely captures both how the MER scientists find a niche in MER and what they are seeking.

Cabrol emphasizes that her personal interest is not to be famous, but to influence the course of science:

> There is the team legacy and there is the direction of research and the perspective you have on something, the personal legacy. Not in terms of "me," with a big M. I don't care about that, rather an intellectual legacy—trying to raise people's awareness about questions. Perhaps other people you share that with will provide their own perspective and augment that vision and raise the awareness a step higher. These are things you need—this little space of isolation, the loner part—to deal with. And then of course you need to go back to the team to implement these ideas. You cannot do that alone.

Like Cabrol, Des Marais views leading other expeditions as relevant to his MER work:

> We're studying a microbial community on Earth that's living in a sulfur-rich deposit. It has interesting parallels to the sulfur deposits that are now being discovered on Mars. You could say this work is preparing us to go and maybe get our hands on some of those deposits on Mars with some tools that could do organic analysis and interpret them.

So I try to keep this research going, which is qualitatively different than what we do on this mission.

Working in an academic social setting, Yingst has developed an identity as a professor that she finds rewarding in part because of how it differs from her role on MER: "I'm at a small university. I am proud and pleased to be working mostly with undergraduates, and putting my expertise into a community where I'm not one voice of many. I am somewhat unique and able to give a different perspective than has been in this community before." Even Rice, who "lives for Mars" and is funded to work on an active Mars orbital mission, has many other interests that he brings into missions: "I combine fieldwork from Iceland, Hawaii, and Australia (volcanism, fluvial, glacial, and periglacial investigations) into these projects." His other interests include "Apollo missions, space history, the Chinese space program, and public speaking on all these topics."

These MER scientists are obviously energetic, with intellectual lives that are far from simple or narrowly focused. They are continuing normal professional lives as researchers, with multiple projects and interests both inside and outside of NASA. In a mission setting that tends to blend individual contributions and shun personal credit, they nevertheless have opportunities to control and develop their own projects, only some of which are directly pertinent to MER operations (such as Yingst's "hand lens atlas").

The scientists' motivations appear complex, with multiple dimensions and justifications. The privacy of a retreat to consolidate one's thoughts might be important for sanity and also necessary for detailed analyses. Many scientists also need to support themselves by working on multiple projects, and with the rover's pace, they have plenty of time to pursue other interests. Being grounded or anchored in personally directed work may enhance a scientist's knowledge, field science skills, and so on, defining a niche for future missions. Indeed, personal projects are part of the ecology of science: developing new perspectives and tools is necessary for the community of planetary scientists to develop and extend its reach.

Straddling Unfamiliar and Familiar Worlds

Previously, we considered how the psychological experience of "presence" enables being an agent in a remote place and thus using a robotic laboratory effectively for fieldwork. The personal aspect of scientific work has also been emphasized in Rudwick's analysis of the cognitive experience of fieldwork. In particular, new encounters are experienced against the background of personal knowledge and social relationships.

In his elaborate analysis of presence, Rudwick suggests that people feel more alive when doing fieldwork because participation on the expedition involves both theoretical and social learning. He calls this a simultaneous "double movement in intellectual and personal terms."[3] Scientists are perceiving and thinking about new phenomena while they are finding new ways of relating to people in a new place. By "intellectual," Rudwick means the cognitive (perceptual and conceptual) aspects of the scientific work (figure 3.6). By "personal," Rudwick is referring to the physical and social separation from home. The scientist's sense of "what I am doing now"—the ever-present aspect of identity—must be recoordinated intellectually and socially, and these two aspects often come together in new ways.

Intellectually, the scientist doing fieldwork moves "from familiar features to unfamiliar and then back to familiar." Specifically for MER, martian features are first recognized as potentially interesting (e.g., an outcrop in Eagle crater), because they seem familiar. A puzzle develops as scientists attempt to categorize and explain features (e.g., are the lines we see indicative of deposits from wind or water?). Further data may enable identifying the foreign object or phenomenon in familiar terms (e.g., the wavy lines are consistent with slow-moving water deposits).

Simultaneously, the MER scientists during the nominal mission (the first ninety sols) moved personally by being radically separated from their institutional community and family. Although living in Pasadena on Mars time, and obviously not on Mars, they worked in isolation behind darkened windows with dozens of people having different technical interests, capabilities, and ways of working.

Rudwick says that expeditions are so fruitful because of such intellectual and personal contextual shifts. He believes it is more than "the perceptual impact of unfamiliar features." Rather, a heightened, directed awareness of the environment, promoting creative work, occurs by virtue of the "social and psychological features of the process of traveling itself." He calls this state of consciousness a "liminal experience," named after the Latin "limen," meaning "threshold." In psychology, "limen" refers to the threshold of perceptual awareness or more generally the start of a transformation or a system boundary crossing: "*Liminality* is that which raises the interaction between trained experience and unfamiliar features above the level of 'everyday' reality. It gives the unfamiliar a 'sacred' aura, in which its deeper significance may be more readily perceived."[4]

The reverential tone of the Haughton discussion about "favorite places" at the start of this chapter illustrates how an expedition in that landscape promotes liminal experiences. Such perceptual sensitivity occurs in part by extension in the felt relationships among the expedition members. The relation of the familiar and unfamiliar may promote openness to learning: "The relation of the pilgrim to his home displays a polarity between fixity and travel, between the secular and the sacred, between what Turner calls the 'structure' or relative rigidity of normal social life and what he terms the 'communitas' or freer social relationships that develop among pilgrims."[5]

Modern expedition participants, like the travelers to Haughton Crater, are familiar with this feeling and the romanticization of the place, the experience, and the group that develops. Surrounded by unnamed features, we share the experience of segmenting and interpreting the landscape—a unifying activity throughout the MER mission.

Further, on expeditions, idealization of relationships develops as a felt camaraderie. We eat and work with the same people every day and encounter no one else, forming our own plans and habits from the relatively unstructured workday that revolves around the place, people, and the technologies we have brought. Similar glowing appraisals can be found in how the MER scientists and engineers talk about each other, which relates as well to their shared history in constructing mutual roles during training, field trials, and the cycle of daily commanding.

Rudwick argues that the combination of intellectual and personal detachment from habitual settings promotes more open interpretation, providing time for fragile new insights to develop without being harmed by "the destructive criticism of other geologists."[6] Of course in MER, the scientists remained in contact with their home communities by email and phone; later in the mission, many returned to institutional business and other projects between MER telecons. But as Rudwick mentions, what matters—at least early in the mission—is not total isolation, but friendly exchanges with outsiders: "Letters … provided an important medium for tentative formulations of new concepts. But this was correspondence with colleagues chosen for their sympathetic receptivity, not the critical and competitive scientific community as a whole."[7]

Of course, the MER project's commitment to post all raw images on the web and to regularly publicize status and results is quite different from the social isolation of a nineteenth-century expedition (as well as all missions before Pathfinder[8]). And although the JPL and the Pasadena environment may not have been entirely like home, the freeways, apartments, and restaurants all provided a largely familiar way of life, quite unlike the outdoor adventure of an actual expedition.

Nevertheless, Rudwick's analysis of premodern fieldwork still has some relations to MER, suggesting another way to evaluate the advantages of bringing the science team together in Pasadena and living on Mars time. Rudwick explains that the personal isolation is a means of "altering the psychological conditions of the geologist's thinking."[9] We can extend this idea beyond geology to include all disciplines of the MER science team. Perhaps travel and isolation in Pasadena served to instill the "aura" of a new place, which proper interpretation requires. For example, the layers we see in Meridiani craters are apparently billions of years old, perhaps two orders of magnitude older than wind or water layers we typically might encounter on Earth. Understanding this foreign planet requires personally "living in the unfamiliar"—physically, socially, and intellectually. And so the drawn shades in Pasadena contributed to the virtual presence that working on Mars required (figure 1.1). Of course, it helped to see the charts, too.

Rudwick remarked that studying photographs was impoverished relative to actually traveling to the place being studied. But he did not foresee how an expedition could voyage on a rover, and the scientists could thus move in their imagination of Mars both physically and intellectually through visualizations. Even with the team distributed and operations reduced over the years, photographs draw the team together, locating them squarely on Mars. Confronting such a landscape (figure 9.1), we stand on a threshold, looking out perceptually into a new space that is also on the boundary of our experience and knowledge. It is this experience of being on the verge of moving intellectually and personally, a threshold of knowledge and identity, that Rudwick called the experience of liminality.

FIGURE 9.1
Meridiani Planum. Opportunity paused on sol 1930 about one-fifth of the way through a 21 km traverse from Victoria to Endeavour crater. *Image credit:* NASA/JPL. *Source:* JPL Mars Exploration Rover Mission, "Opportunity: Navigation Camera: Sol 1930," http://marsrovers.jpl.nasa.gov/gallery/all/1/n/1930/1N299526557EFFA3PEP0743R0M1.JPG (accessed July 1, 2009).

With a magnificent effect that would probably surprise Rudwick, given his focus on actual travel, MER photographs stimulate personal and public excitement, providing an aesthetic experience that "is the best next thing to being there." In subsequent sections, I consider the aesthetic aspect of MER photographs and their effect on the public identity of the mission.

The Public Persona

On top of the demands of multiple shifts, living on Mars time, and juggling two rovers on opposite sides of the planet, the mission had from the start a public face that required further adaptation by the scientists and regulation of their activities. The mission was in fact *deliberately public*. The unprecedented response to Pathfinder's website in 1997 had shown everyone that it was now possible to engage the public by continuously providing images from Mars. The remarkable images made the mission visually concrete for the public, and the effect "stunned the Net world."[10] Without images, the operations would be abstract and the reality of being on Mars poorly imagined, if considered at all.

On earlier NASA missions, making science data available to the public was not easy, and the real issue was whether to share it even with other members of the science team. In particular, Bell was profoundly influenced by his experience as a student on Voyager: "I was not prepared to see the greed, secrecy, anger, and just plain nastiness some scientists exhibited concerning access to, sharing of, and publication of their data." Squyres agreed, having had similar experiences, and "most of the rest of the rover science team felt similarly." So it was decided that all images would be made public "with no restrictions, embargoes, or proprietary data periods."[11] Thus, ironically, a technology that was evoked in the name of "public outreach" and that made sharing with anyone easy also promoted a collegial, collaborative spirit in the mission team itself.

The MER scientists were trained on how to provide "outreach" as a team as an integral part of surface operations. In particular, during the August 2002 FIDO field test, procedures for interacting with the MER team were enacted during simulated surface operations. During this time, the team established a decision-making and approval process for daily releases of two captioned images and rover location updates, held a practice press conference, involved graduate student interns in the science process, did a live webcast, inaugurated the MER website, and contributed to television and cinema documentaries.[12]

Throughout the mission, the rovers' activities have been documented on the JPL website (at first every sol and then varying over the years to about once a week) with representative images and explanations of their significance. On selecting the first photographs of the lander, Squyres said, "We want to make a good first impression on the world." Accordingly, the team participated in press conference panels, magazine interviews, and television shows (figure 9.2/plate 22) and even starred in an IMAX movie with Paul Newman.[13]

FIGURE 9.2/PLATE 22
NBC television crew filming the science
assessment room as Squyres presents
(February 12, 2004).

Perhaps some of the scientists' interest in sharing the excitement of the mission comes from the influence of Carl Sagan, whose space science books, television show, and public appearances are very well known. For Squyres in particular, "Carl was a kind of teacher, advisor, mentor through graduate school. We did some science together and I also gained from him an appreciation of the importance of being able to, trying to, communicate the results of our work to the people who paid for it."

As if not realizing the effect of the mission's comprehensive efforts to engage the public, the scientists were sometimes bewildered how everyone knew them and were eager to discuss the mission. Carr relates his public experience:

> It's incredibly public. When I was flying back and forth to JPL and I'd be sitting in the airport, waiting for the plane, and people would sit next to you and you'd get talking, and people would realize: "My god! You're running those rovers!" *(animated and happy)* How incredibly informed everybody was, it was amazing! Just amazing, everybody was involved. I'd go to the office here, people would stop me. I'd go to my horse people here in Woodside [California] and they would ask, "What's going on?"

Thus another aspect of the scientists' identity, inherited from the broad public interest in the rovers, is having a job that everybody knows about. Rice said, "When I meet somebody and they ask me what I do, I say, 'Astrogeologist,' and almost immediately I say, 'I'm working on the Mars rovers.' Because everybody knows what that is."

Des Marais recognized how promoting the mission has provided another perspective on publication—they are influencing the scientists and engineers of the future (figure 9.3/ plate 23): "[We are] publishing the exploration experience and findings to the outside community. And it's not just because that's a nice egalitarian thing to do. We are actually recruiting the kids, the people who are going to kick the ball through the goal, down the road, in finding life on Mars. It's these kids that we're reaching by education and outreach." Squyres relates how working with students is "tremendously valuable, and very rewarding":

> The students are a huge asset. They bring energy and enthusiasm. They're willing to work *insane* hours for almost no money, just for the excitement and the privilege of being here. The students have been absolutely fundamental to what we do. There are dozens of students now who grew up on MER, learned flight operations on MER, and they're the guys who are going to be doing the spectacular Mars missions and Europa missions, twenty, thirty years out. They cut their teeth on MER.

Squyres remembers how his own involvement in the space program began in the very same building at JPL decades earlier (where the MER project is located and where our interview occurred):

> When I was in grad school, I got to work as a student on the Voyager mission to Jupiter and Saturn. You know, and I was here in *this very building*, one floor down, third floor, in 264. In 1979, I was twenty-three years old, working with the Voyager imaging team when we discovered the volcanoes on Io, and the ring around Jupiter, and that maybe there's an ocean on Europa, grooved terrain on Ganymede..... And I was just a kid, you know?

I can never repay the Voyager project and the Voyager imaging project for the life-changing experience that mission gave me. But what I can do is help pass it on to some other students. And I hope they do the same.

At Haughton in 1998, Jim Rice left a deep impression by telling me, "If I can't go to Mars, I hope I can train the people who do go." He related how training geologists at an analog site, as astronauts were trained for Apollo, is "the next best thing to being there."

Similarly, in considering the educational effect, Cabrol is thrilled how they convey the mission's adventure on a global scale:

> Five hundred years ago, you would take a boat and discover another place. And you would see that. You would record that, come back and tell your story to a limited number of people. Of course you didn't have web access, you didn't have telephones, so the word was spreading at the speed of the horse or human voice, about exploration. Today, six billion people on January 4, 2004 discovered a new site on another planet. This is human exploration. Not human because humans were there, but because we were *all* there, together, through a robot!

Thus, the rover and web technologies enable people to identify with the mission and to experience exploring Mars. Nicks suggests that public interest in robotic missions has occurred throughout the space program: "Perhaps one of the most significant observations, made by a number of writers, is the fact that the adventures of our automated spacecraft have been enjoyed and shared with mankind in a real-time drama, literally unfolding before millions of eyes."[14]

Writing in 1985, Nicks could hardly have imagined what kind of "immediacy of sharing" would be possible with the Internet, famously exploited for the Pathfinder mission. Now with email added to the mix, Cabrol receives requests and is touched by other families all around the world:

FIGURE 9.3/PLATE 23
The Red Rover student team examines the Eagle crater panorama, Meridiani science assessment room (February 18, 2004). *Note:* The students in the photograph are identified as Kristyn Rodzinyak and Cheng-Tao Chung in Kristyn's blog, "Projects: Red Rover Goes to Mars," the Planetary Society, http://www.planetary.org/programs/projects/red_rover_goes_to_mars/journal_kristyn.html (accessed October 4, 2006).

People are asking questions because they were curious about data they would see on the Web, and asking me if they were right or wrong, and where I am heading. These were people asking just to share a moment, to share something special. So I would send a photo with a little note, personalized. One person was asking if he could have a photo for his children, who were really small but in a few years from now would be able to understand its meaning. So see, this is something that people want to share over generations.

Just as a media celebrity would send an autographed photo of him or herself, Cabrol sends photos taken by MER, on which she writes a personal note—a mark of the image's authenticity and value and thus her identity as a scientific explorer.

Educational animations and videos are posted on the MER website, including the breathtaking descent to landing sequence and serendipitous video of the dust devils. Trebi-Ollennu related a humorous incident in which an engineering mistake produced a fascinating and educational result. By inadvertently using a variable to control an MI sequence that also indicated whether the rover was driving, they recorded an animation that "turned out to be great for outreach."[15]

Presenting a public face for the mission has not always been so easy. Most notably, although Squyres could exclude the media from the end-of-sol meetings, he was obligated to participate in press conferences, which did not respect that the scientists were living on Mars time: "As long as all I had to do were Mars-like operations, I was fine. It was difficult when people would call a press conference in the middle of the night. *(laughs)* My night! There were times when I would go home, I would sleep for four hours, set my alarm to wake me up at midnight, my time, so I could come into JPL, do a press conference and go back to sleep." Squyres explains that this makes sense "because the reporters happen to live on this planet!" It might be more dramatic to call the press conference at 3:00 a.m., but the press would not show up. At other times, when the planets were aligned, Squyres' fate was more joyous: "Several times when we were turned around 180 degrees from Earth time, somebody decided to have a NASA Headquarters press conference. So now I've got to fly across the country. I took red-eyes—middle of the day for me! I've got my light on, I'm drinking coffee, working on my laptop. Everybody's glaring at me! But that worked just fine."

Aesthetic Experience: Beautiful Images and Revered Places

Many people are captivated by the aesthetics of the martian surface. The lead scientist for Pancam, Jim Bell, assembled a book of photographs, *Postcards from Mars: The First Photographer on the Red Planet*. The book is strikingly personal, with a subtitle crediting Bell as the person who took the photographs.

Of course, Bell could not alone decide which images would be taken or personally command the rover. Turning upside-down the tales about the "hard-working rovers," he lauds "the tireless team of scientists and engineers who control the rovers…. [They] have the privilege and quiet pride of being the 'cameramen' behind the incredible color pictures being shot by *Spirit* and *Opportunity*."[16] But describing the working relationship of people and the rovers can be tricky: he says that the rovers "have allowed us to be, in a sense, the first 'photographers' on the Red Planet," placing the quotation marks oddly—the team members are in fact photographers, but they are not "on" the Red Planet.

In designing a photographic sequence (MER Observations), Bell was well aware of the aesthetics, at first in framing the images and later in image processing by selecting special, interpretive effects. With the flexibility of MER's cameras and the sheer number of photographs involved, instead of merely "acquiring images" as on Viking, he was *taking photographs*—by including foreground for depth, balancing sky and ground, using filters, and postprocessing in the darkroom. Consequently, he concludes, "The artistic, aesthetic—photographic—aspects of these images are my doing."[17] Here is the personal scientist, one making inherently scientific products into works of art.

The intentional aspects are pervasive, but nevertheless serendipity is familiar in art, too. The difficulties of the engineering process introduce the possibility of unexpected results, like Trebi-Ollennu's educationally useful MI mosaic mentioned previously. In fact, the best view from Husband Hill (figure 9.4/plate 24), one of the mission's most celebrated images, was according to Trebi-Ollennu "an absolute mistake." It was a Navcam panorama that was supposed to include the IDD work area. Very late in the planning process, the rover planners included a command to turn the rover for better communication of data. But they didn't tell the Navcam PUL, who inadvertently produced this image with the rover turned around 180 degrees. When they put the panorama together, they were surprised: "There is Spirit! On the summit of Husband Hill. It's the whole rover, with this vista in the background, there's a dust devil dancing through the scene. It was this fabulous picture that we'd never intended to take!" Indeed, when Squyres reviewed MER pictures exhibited at the art museum at Cornell, he "was struck by how many of them were accidental, the consequence of sequencing errors, mistakes, things that we did wrong and we got something unanticipated."

On the other hand, how much flexibility do the scientists have to specifically reposition the rover for visual effects? Squyres explains that scientific responsibility includes not only relating to the scientific community, but also relating to the interests of the public:

> While fundamentally, we feel that we are there to do scientific research, we are humans privileged to do an extraordinary job and we feel very much that we're doing it on behalf, not of just our scientific colleagues and peers who aren't on the team, but on behalf of all the millions of people who've paid for, enabled, and followed this mission. And they don't just care about what's the silica content of this particular outcrop. They want to see Mars in all of its seasons, and all of its moods, and with all of its beauty. And we do take images that are just to be beautiful.

At this point, we can feel with Squyres the commonsense insistence that after their work was done, it was okay to do something for fun that everyone would appreciate. For

FIGURE 9.4/PLATE 24
Spirit on the summit of Husband Hill; as Squyres explains, although
studying the outcrop was essential, the image itself had no initial
scientific justification. *Image credit:* Craig Covault, "Rover Atop Martian
Mountain Pushed to New Limits for Critical Science Data," *Aviation
Week* (November 13, 2005), http://www.aviationweek.com/aw/generic/
story_generic.jsp?channel=awst&id=news/11145p1.xml (accessed
January 15, 2008).

example, the initial panorama at the top of Husband Hill revealed a steep outcrop they called Hillary, which they chose to climb to take a panorama:

> The science was on that steep face. We drove up to the base of it, and we did a bunch of IDD work, and we had done the science that we had come there to do. We spent *three sols* driving around *to the very top* to the point where the Pancam cameras were the highest thing on Husband Hill, and we took a full 360-degree panorama. What was the science in that? There was none. Okay, but if we'd just climbed the first mountain on Mars, are you going to tell me that I'm not going to the very summit and take a picture? So yeah, there are times that we take pictures just because they're cool.

Here he echoes the mountain-climbing ethic "just because they're there."

Why Are the Images "Beautiful"?

So what do Bell and Squyres mean by "beautiful images"? Whether pondering press release images (figure 5.4) or our own favorite discoveries in the web collection (figure 9.1, described by Sims as "gorgeous"), we can experience entering into the space, drawn within and feeling we know these sights as places we could go. Many of the photographs present a desolation that evokes the open, untouched geology of American southwest deserts. Losing sight of the rover and scientists, we sense the cold light falling on pristinely patterned sands in an ancient, apparently lifeless silence. This barren world, visually occupied as a whole, transcends argument and interpretation. We stand alone in the open realm of a largely unknown planet, the first to ever gaze upon it.

What accounts for our attraction to such landscapes? Perhaps part of the allure of Mars is its perceived innocence, its pure and unspoiled nature, away from the complexity of Earth. Related romantic ideals appear frequently in sociopolitical statements about the space program. For example, the International Space University declares the objective "of a peaceful, prosperous, and boundless future through the study, exploration and development of Space for the benefit of all humanity." In chapter 2, we contrasted this view of "exploration" with a cognitive analysis. In this section, we consider how the activities of art, science, and travel are mutually influenced during the MER mission, such that emotional experience provides a foundation for scientific work, and how presenting the work artistically emotionally engages the public and bolsters support for the mission.

In relating geographic exploration and art, we might begin with the eighteenth-century European voyages around the Earth. These early expeditions are reinforced in Western cultural memory through the beautiful botanical drawings and the images of the artists who recorded and idealized the foreign places and people. For example, William Hodges, an English painter (1744–1797), sailed on Cook's second voyage to the Pacific and is "best known for the sketches and paintings of locations he visited on that voyage, including Table Bay, Tahiti, Easter Island, and the Antarctic."[18] Hodges's artwork, in combination with the geographic and scientific knowledge the voyages returned, "transformed the way that Europeans looked at the world beyond Europe."[19] In conveying exotic beauty, Hodges's art romanticized the lands being explored, idealizing them as pure and unspoiled places, as in the painting *Tahiti Revisited* (1776).

Early scientific explorer Alexander von Humboldt found within Cook's voyages a pleasing combination of scientific accomplishment, aesthetic sensitivity, and literary skills, which he enthusiastically adopted.[20] The German Romantic movement holistically embraced such grand themes, which Humboldt's ambitious work reflected: "the painstaking empiricism, the remarkable breadth of intellectual interest, the passion for the beauties of nature, and the commitment to a universal science." Accordingly, Humboldt believed that aesthetics plays an important role within natural inquiry: "Man's aesthetic sensitivities could, if suitably trained and applied, transcend the limitations of reason, penetrate beyond surface phenomena and, sensuously and intuitively, grasp the underlying unities of nature."[21]

To tie these themes to MER, we need first to distinguish between romanticizing and the Romantic Movement. A romantic attitude is a general "imaginative or emotional appeal of the heroic, adventurous, remote, mysterious, or idealized characteristics of things, places, people."[22] The Romantic movement (or Romanticism) was an eighteenth- to mid-nineteenth-century reaction against the idea that all knowledge comes from the senses and reason; it emphasized the role of imagination, emotion, and personal expression. Accordingly, a Romantic literary style could be autobiographical and introspective, and subjects of interest included "an exaltation of the primitive and the common man, an appreciation and often a worship of external nature, an interest in the remote in time and space, a predilection for melancholy."[23]

MER imagery relates to both romanticization and the Romantic movement. First, like the later German Romantic movement, the interests of the scientists in presenting MER images and, more broadly, my presentation of their work is not to react against rationalism, but to complement it by showing how aesthetics contributes to the scientific inquiry. Put another way, this perspective claims that the rational work is motivated by more than "curiosity" or fame. Accordingly, when MER scientists say that work on the mission is sustained because it is fun, they are adopting the Romantic perspective. Put more starkly, the team's adherence to the "scientific method" does not fully explain the quality or extent of the work they have accomplished—their emotions and aesthetics were necessary, too.

Second, Jim Bell's aesthetically motivated book *Postcards from Mars* fully embraces Romanticism's style of individualized expression (in which the team labors, not the rovers), including introspective autobiographical material and an appreciation of a remote, external nature. The empty spaces of MER images lend themselves well to melancholy poetic tributes when the rovers die.

Third, contrasting with Bell's first-person narrative, the typical "rover as hero" third-person perspective is also plainly a romanticization, which—as mentioned previously—has rhetorical, poetic, and even team-building motives.

Finally, my analysis of the MER scientists partly adopts a Romantic literary style, elevating the individual's experience, showing their personal challenges and exaltations through autobiographical anecdotes that stress their emotions and imagination.

In short, the motives and methods of art and science are complexly related, providing a vibrant topic for philosophers and introspective scientists. Such sensibility is expressed by Freeman Dyson in *The Scientist as Rebel*, which praises the heroic intellectual who engages in creative struggles, aiming not to win control but as counterpoint to prevailing cultural veils and values, to focus on spiritual eternalities: "The chief reward for being a scientist is not the power and the money but the chance of catching a glimpse of the transcendent beauty of nature."[24] In this view, perceived beauty in nature relates to the beauty of "constructive science," the realm of ideas. That is, scientific aesthetics combine physical and explanatory, logical beauty. Indeed, he goes so far as to say, "Science is an art form, not a philosophical method."[25] Accordingly, Tufte's *Beautiful Evidence* argues that aesthetic presentation of data is rich with informative relations, both perspicuous and economical. MER route maps demonstrate this style (e.g., figure 2.1).

If you ask a layperson what they know about Hubble, "beautiful images" is likely to be mentioned first. This is the first impression of many scientists in planetary missions, too, as Chaikin relates about Viking: "Backlit by the morning sun, the scene was not only superbly detailed, it was undeniably majestic…. Viking 1's cameras had done something that wasn't in their design specs: They'd revealed the beauty of Mars." And referring to the subsequent sunset photo taken August 30, 1976: "It was something only an explorer could have discovered: Martian sunsets are blue."[26]

For scientists, beautiful images impress and intrigue. Mars may appear more monotonous to people who are not engaged in inspecting and moving through the terrain, who lack the visual language of geology. They see only lines in Victoria crater, for example, which scientists moved in to investigate (finding them to be bands of wind-borne sedimentary deposits later pervaded by briny water). This parsing of the terrain may make it more interesting and beautiful in the imagination (figure 9.1): do you see ripples of sand or feel the directionally blowing wind?

Pilgrimages to Revered Places

In his analysis of the "liminality" of fieldwork mentioned earlier, Rudwick mentions how a deeper significance develops that beckons repeat journeys: dwelling in the tacitly perceived "aura" of a place makes it revered; it becomes "a spirit place."[27] Kohler, citing Rudwick, says the urging call of a place continues today: "Field geologists experience expeditions as analogous to religious pilgrimages."[28] Perhaps surprisingly, some of the aspects of nineteenth-century expeditions remain where travel is "slow, expensive, uncomfortable, and even hazardous."[29] If you have engaged in the weeks of preparation, the multiple-day journey, and the harshness of the isolated Haughton–Mars camp on Devon Island or similar places, you may have experienced the feeling of making a pilgrimage. Then, for years after, you feel an aching desire to return. Starting in 1997, a group from Johnson Space Center traveled to the area around Flagstaff, Arizona, in the first two weeks of September, calling their expedition, "Desert Research and Technology Studies" (Desert RATS or D-RATS). In practice, this ritual, which continued for more than a decade, felt very much like a retreat for engineers. Its aura was reinforced by the long hours of work and tribulations of an often-late "southwest monsoon" that left a seemingly limitless horizon of red mud dotted with freshly greening sage.

Thinking particularly about the four-year trek with Opportunity, the historic stops and investigations, the thousand sols of painstaking choices, one could feel in 2008 the anxious pleasure of another multiple-year traverse to distant Endeavour crater. Out here, our survival unknown and success questionable, we wondered what further discoveries await:

> The pilgrim's travel toward his goal is a route of every more sacred character: his sense of expectation, and hence his receptivity to new understanding, is progressively heightened.... The geologist's goal is "sacred" in the sense that some deeper understanding is anticipated from exposure to the geological features that constitute that goal, just as the pilgrim hopes to experience some "theophany" or disclosure of the divine at the focal point of his pilgrimage.[30]

Rudwick concludes that the experience of liminality is rare on modern field expeditions, "except perhaps when the excursion is long and leisurely, and the party is small."[31] By design, analog Mars expeditions to places like the Canadian Arctic, the American southwest desert, or the high Andes seem to fit, and certainly these first overland expeditions on Mars have evoked feelings of reverence. Perhaps not coincidentally, at the mission's conclusion Squyres described the view of Home Plate from Husband Hill as "what turned out to be for Spirit the geologic promised land."[32]

Having considered the origins and effects of aesthetic experience in scientific exploration, we return finally to the sharing of images with the public and the effect these images have on people who have not directly participated in the scientific work. In *The Art of Travel*, Alain De Botton provides a useful perspective: "And insofar as we travel in search of beauty, works of art may in small ways start to influence where we would like to travel to."[33] Representation of the scientists' own sense of charm and wonder—their idealization of the land and their romantic view of their scientific exploration enterprise as reflected in their framing, selection, and color manipulation of photographs—promotes and taps into the same feelings in the public.

De Botton tells us about the influence of art on the public's interest to travel, including Hodges's paintings, as well as romanticized renditions of England's Lake Country in the second half of the nineteenth century. Tourists motivated by these images sought these places "in an attempt to restore health to their bodies, and more important, harmony to their souls." They came by horse and carriage, fed as well by the poetry of Wordsworth and others—the tourists "sought out … the hillsides and lakeshores … whose power he

CHAPTER 9

had described in verse…. By 1845, it was estimated that there were more tourists in the Lake District than there were sheep."[34] Now, looking again at Bell's *Postcards from Mars*, you might reconsider what effect such a book might have.[35]

Emotional Experiences

The scientists' recognition and collection of MER photographs is just one way in which scientific work and emotions are intertwined. Considering now other emotional experiences during the mission, we find a pattern in which certain moments are unforgettable and thus provide another window into how deeply people relate to the Mars rovers and the mission in ways that transcend purely scientific interests.

On January 4, 2004, I watched Spirit land on NASA TV, viewing the full-screen Internet image on my laptop as I sat in an Orlando hotel room. I experienced overwhelming joy, which brought tears to my eyes, feeling a great resolution after the long wait for success, realizing that we had avoided a horrible crash like Mars Polar Lander in 1999 or Europe's Beagle lander in December a few weeks earlier. Squyres reported a similar incident: "Jim [Bell],… slumped in a chair … there were tears in his eyes…. He looks up at me. 'It works, man…. It works.'" Then later, back in his hotel room, Squyres reported that he tried to "be cool" but the emotion was overwhelming: "This is so good. I can't believe how good this feels…. Pancam is really on Mars after all these years. The whole damned thing is on Mars. I dissolve into tears."[36] This emotional experience at the signal of mission success after months of suspense may be common. Bill O'Neill, former project manager of Galileo, reported, "Some of the senior engineers in the Probe area, when the signal came to say that everything worked, they were actually crying."[37]

Rice, Sims, and Cabrol each told long detailed stories about the night Spirit landed on Mars. Each felt a strong historic and personal connection. Rice said: "I remember before Spirit landed, I went out and looked at Mars and was thinking, Mars is never going to be the same again, after tonight, one way or another. It's not going to be the same, because we're going to land there or we're not going to land there. If it crashes, it'll be the frustrations and disappointments. And if it works, it'll be unbridled joy." After saying how he didn't feel particularly overwhelmed emotionally, Sims relates an unmistakably exciting story:

> The real nail-biting moment was the landing of Spirit. Okay, it landed…. We think we hear a bounce. Nope, we're hearing random bounces…. We're sitting there in the SOWG room, sitting around the SOWG room and they're broadcasting signals…. They're coming down, and there were long moments…. Because Rob [Manning] admits he didn't know what was happening, and he was trying to figure what was going on, and it was actually bouncing…. It was like, you get this high … "Great! We got a signal." And a long time without anything … is that okay? "Okay, another one!"

Obviously, the suspense and the great importance of every event during the descent and landing kept people in rapt, sharply focused attention. Cabrol remembers:

I thought Spirit was going to be okay, so I was not surprised when everything was going *(snaps fingers)* like clockwork. Parachute deploying *(snaps fingers)* … radar … And then, when Spirit stopped talking, I didn't bulge. I was not worried at all. I was not worried for 14 minutes and 59 seconds. On the fifteenth minute, I thought to myself, "We didn't lose it now." I was not done phrasing that when I heard, "We hear it!" *(very happy laugh)* It was just at that point we jumped all over the place. It was an incredible ride, an incredible ride. Very personal. There is the team aspect, of course. But yes, it has been a personal journey. Very personal journey.

Strikingly, Squyres concludes his book by saying, "I love Spirit and Opportunity,"[38] explaining that one should not say such things lightly about machines, but this is what he feels. Rice anticipates how he will feel when the rovers die: "These things have done some great work and allowed us to really view Mars in a way we never had until now. There's an emotional attachment to them. As I think everybody has said, it's really going to be a sad day when they're gone."[39]

10

The Future of Planetary Surface Exploration

What We've Learned: Communal, Public, Virtual Science

What have we learned about using a robotic laboratory to conduct field science on a planetary surface? What are the lessons for the future scientific exploration of Mars? More generally, what have we learned about the design of operations in which routine, programmable activities (such as a Mini-TES scan, moving to another site, and even finding rocks of a certain type) can be relegated to robotic systems?

I have sought to answer these questions by examining what accounts for the quality of the MER scientists' work, focusing on their direct experience: in investigating the martian landscape, in working as a team, in their public relations, and in professional activities and emotional moments. The MER mission—as a combination of robotic and software technology, "one instrument, one team" organization, bonded over ninety sols of Mars time and then distributed for more than eight years—constitutes a new kind of exploration system, providing a prolonged virtual presence for a large team of scientists roaming through and studying a landscape of plains, hills, and craters on a scale ranging from molecular analysis to rocks and layered deposits to dust devils and clouds. Understanding the MER exploration system's capabilities and limitations requires considering how people adapted to participating on this new kind of expedition—offering presence yet anonymity, daily engagement at a glacial pace, and laboratory analysis in an exploratory context—an inquiry in which scientific systematicity plays itself out within a realm of opportunism (figure 4.3). Not the least of the surprises is that robots with a ninety-sol warranty have been used productively for so many years. And not the least of the difficulties in learning from MER is how the fundamental achievement of providing agency to the scientists to observe and manipulate the martian surface has been obscured by the portrayal of an independently minded "robotic geologist," yet this metaphor had considerable value in designing the rover and science operations.

This study began by asking how working remotely through the technology of a rover affects how the scientists see themselves, their work, and what makes them professionals. I especially considered philosophical, social, and psychological issues related to identity, such as relationship of the rover to their bodies, referred to informally as mediating the "analog" and "digital" worlds. Here is a summary of what the scientists' experience and orientation have revealed about the MER mission.

Most fundamentally, by "science operations," we mean particularly the scientific fieldwork occurring on the planetary surface—what the scientists are doing on Mars—and not just the decision making and engineering occurring on Earth. Such focus on surface manipulations and sampling was natural for Apollo, but is different from the mentality of operating a lander like Viking, and is even more unlike remote scanning from planetary orbiters like Mars Global Surveyor and flybys like Voyager. MER has provided the ability to systematically investigate kilometers of terrain and drive into craters in the changing seasons of Mars—a somewhat unexpected possibility that required the team to construct an organization, tools, and protocols that enabled laboratory and field scientists to explore the landscape while the engineers improvised workarounds for failures and kept the rovers safe.

Cleverly and carefully exploiting their shared and limited technological resources, hundreds of scientists and engineers moved together collaboratively as field and laboratory disciplines found new ways to cooperatively study the martian surface. Within the context of communal, public, and virtual experience, MER operations have challenged and refined scientific practices, altering the scientists' identity—indeed, as Cabrol emphasized, the mission affects how billions of people around the world see themselves. For the MER scientists, being a member of the team provides an opportunity to experience being on Mars and thus realize their personal dreams of being explorers.

Yet although telerobotics made the investigation possible, it distances the scientists from a landscape they would prefer to walk through, and perhaps through the very success of virtual presence offered by the tools and their imaginations, the afforded agency and resulting pace are made tolerable but not satisfying. Further, the necessary communal approach requires that individuals shape the investigation without controlling it—on the team, they must be agreeable yet prodding and must compromise yet speak strongly among strong proponents for disciplinary points of view. The scientists become for the large part anonymous in the daily accomplishments of the mission. The press reports speak of "Opportunity's investigation" and will never mention an observation suggested by Cabrol, Carr, Des Marais, Rice, Sims, Squyres, Yingst, or dozens of other scientists, or mention how Matijevic, Trebi-Ollennu, or dozens of other engineers made possible these actions on Mars.

When asked to fill in the blank, "Working with a rover is … " the interviewed scientists and engineers responded, "fun," "a joy," "a symbiosis," "equivalent to working with a team," "frustrating," "humbling," and "a dream come true—the next best thing to being there." Based on the elaborations and qualifications, it seems likely that everyone would agree with all of these perspectives. The overall tone is positive but qualified by an awareness of limitations and ever-present vigilance and responsibility imposed by the work.

Joined at the hip but advocates for their own particular disciplines, valued for their place on the team but tolerating the submersion of identity because it is equally shared and because they are able to express themselves through the rover, this technologically mediated field science created new scientists. Through the rover, they forged a team that explores Mars as a group, coaxing their laboratories through treacherous Martian sand and steep rocky craters. Earlier technologies—some orbiting silently above a planet, some simply planted on the surface—gave scientists months or even years to consider and formulate plans. MER, with its two boots on the ground, demands a different kind of thinking: a thinking-in-place focused on a few meters ahead or the next few days.

"Doing science" devolves into teleconferences, image manipulation, and computer analyses that provide their only contact with—yet further alienate the scientists from—the rocks and chemistry of Mars. Nonetheless, each person, in discovering a way of participating in relation to the robotic technology and its products, finds a unique place in this bigger picture. Each person interviewed, from the youngest to the oldest, regardless of his or her area of specialization, expressed clearly the importance of finding a niche, of being useful. This self-directed matching is inherently reflective, active, and ongoing: what can I do here? What is my capability? Where can I make a contribution? Carr's experience shows how careers are defined by missions (stated so well by Cabrol), but new technology challenges the scientists' ability to relate and contribute. Sims tells us conversely how being on a mission is perhaps not the best way to advance an engineer's career, but—as for the scientists—it fulfills his personal commitment to be an explorer. Then coming from the other direction, Rice's personal project of searching every nook and cranny of images for clues shows one way of realizing a passion for exploration, despite not being on Mars, while reinforcing his individuality within the science team.

Sometimes being productive is marked by a unique contribution—the first person to see the ripples in Eagle crater, the first to understand a rover fault, the first photographer on Mars—but usually it means just adding useful information or working with others. Each person experiences this process of being productive relative to the rover and the scientific mission as a social responsibility, a matter of personal integrity, and a shared pleasure when the niche is found. Otherwise, the experience, as Carr sometimes discovered, will not be fully satisfying, and another personal relation must be found. And here we saw the irony that upon leaving the team, Carr turned around to write about MER in a revision of his seminal textbook on Mars. Thus each person, in discovering a unique angle, a way of participating in relation to the technology and its products, remakes themselves.

Subsequent sections elaborate how the practice of scientific fieldwork has developed through MER and lessons for organizing people and configuring robots for planetary surface exploration.

A Combination of Scientific Practices

The overarching theme of this book is how investigating a planetary surface through a robotic laboratory changes the practice of field science, and in terms of an "exploration system," how the relation of people and machines enables scientific fieldwork to be conducted remotely. *Practice* is essentially what people do—their activities. Human activities are inherently social (defined within and contributing to group interests and actions), are located in some setting (including distributed meetings), and usually involve tools (ranging from documents to physical instruments), constituting what operations analysts focusing on organizational development call a *sociotechnical system*.[1] We have seen that MER combines field and laboratory science and how it relates to early scientific exploration (e.g., Humboldt) and other planetary missions (table 2.1). Table 10.1 brings these ideas together by characterizing the MER exploration system in terms of the methods and team organizations in scientific projects.

TABLE 10.1

Categorization of scientific practices by sociotechnical organization. Research is often oriented around a setting for applying instruments, in which participation and information is restricted. Project types vary greatly in the number of members, size and complexity of equipment, travel required, and hence cost.

TYPE OF SCIENTIFIC PROJECT	SETTING	TYPICAL DOMAINS	PUBLIC ACCESS	TOOLS	SCALE	NO. OF PEOPLE
FIELD SCIENCE	VARIES: ISOLATED OR MUNDANE	BIOLOGY, GEOLOGY, SOCIAL SCIENCES	CLOSED	HAND (E.G., PEN, LENS, CAMERA)	INDIVIDUALS OR SMALL GROUPS	ONE–FEW
LABORATORY SCIENCE	IN A BUILDING	BIOLOGY, CHEMISTRY, PHYSICS	CLOSED	LABORATORY EQUIPMENT	INDIVIDUALS SHARING	FEW
MODERN EXPEDITION	BASE CAMP; CITY OR EXTREME	ARCHAEOLOGY, GEOLOGY, BIOLOGY, ROBOTICS	VARIES	HAND PLUS SOME LAB EQUIPMENT	MODERATE TO LARGE	FEW–10S
BIG SCIENCE	OFTEN ISOLATED SITE	NUCLEAR PHYSICS, ASTRONOMY, OCEANOGRAPHY	CLOSED	LARGE MACHINES	VERY LARGE	100S–1000S
PLANETARY MISSIONS	EXTREME	PLANETARY SCIENCES (E.G., PHYSICS, GEOLOGY)	OFTEN PUBLIC WITH CLOSED ASPECTS	ROBOTIC LABORATORY, PERSONAL SOFTWARE, PUBLIC WEBSITES	VERY LARGE	1000S

This classification of sociotechnical organizations can be contrasted with historical analyses, which categorize activities by time periods or development phases (e.g., Pyne's "ages of discovery"[2]). "Scientific practices" (or enterprises) are different ways of relating people, tools, and settings in scientific work. They are not phases in the history of science per se, though obviously scientific progress and new technology has made new kinds of organizations and topics possible. In particular, modern expeditions and big science combine and scale up field science and laboratory science, respectively.

The MER exploration system combines aspects of each kind of scientific practice. By thinking about these alternative settings and methods, we can realize the place of the mission in the history of science, as well as the adaptations and challenges imposed on the science team, and possibly generate ideas about the future practice of studying Mars:

1. *Field science* (or fieldwork) involves in situ studies, generally by individuals or small research groups with a lead scientist. The setting is determined by the scientific topic, and today is not necessarily remote or dangerous (e.g., anthropologist's studies of city offices). These projects are almost invariably not public until the studies are peer-reviewed in publications. Tools are generally personally carried by the scientists throughout the daily work (e.g., cameras and computers for logging data).

2. *Laboratory science* emphasizes controlled experimental protocols, usually in a dedicated space, using specially designed instruments. Classic early examples include studies in optics, chemistry, electromagnetism, and other fields.[3]

3. *Modern scientific expeditions* are generally large-group, highly organized fieldwork activities. Scientific expeditions are associated with academic or government research institutions, usually with lead investigators who arrange funding; they are a common part of graduate studies and professional careers. Modern expeditions often work from a base camp for weeks or months, bring laboratory equipment to the field, involve dozens of people, and are expensive.[4] The permanent camps in Antarctica, such as the British Antarctic Survey's Halley Research Station, have been occupied for over fifty years and demonstrate concepts of modular design, assembly, life support, recycling, and international participation relevant to a planetary outpost.[5]

4. *Big science* is a name generally associated with large instruments and machinery (e.g., particle accelerators, telescopes), with projects lasting decades and budgets ranging from several hundred million to billions of dollars, constituting a scaled-up version of laboratory science. These projects tend to be managed by permanent institutions and may employ dozens or even hundreds of international scientists working on theoretically related individual and group investigations.

5. *Planetary missions* use spacecraft to transport and control instruments in a form of telescience. These are often big science projects involving a highly intermingled system of fundamental science, engineering, and computer science.[6] Unlike other big science efforts, the management of planetary investigations is organized around the concept of a bounded *mission* rather than a permanent institution.

MER combines field science with laboratory science on a kind of expedition. MER is also a kind of big science, a significant national program and investment with international partners.

Undersea archeology using teleoperated robotic submersibles provides an Earth-based analog of planetary missions like MER. For example, the field science expeditions using Jason described by Robert Ballard are multidisciplinary (involving archeologists, oceanographers, and ocean engineers) and use advanced scientific instruments and support facilities (e.g., a nuclear research submarine).[7] Such investigations straddle categories of laboratory and field science, but—although expensive and requiring sophisticated planning and organization—do not approach the level of funding or the extent and complexity of design, development, testing, and operations required for planetary science. In particular, a remotely operated vehicle like Jason need not be programmed to carry out daily operations, does not navigate or filter data autonomously, and is not kept operating in a single underwater landscape continuously for a multiyear investigation. Nevertheless, the analogies relating field science and robotic systems are strong, and much might be learned by studying further the relation of undersea exploration to planetary missions.[8]

In contrast with planetary landers, orbiters, and flyby missions, the MER exploration system is designed to enable field science. However, many of the rover's instruments are not used by Earth field scientists in situ, requiring them to relate and adapt field and laboratory practices on Mars. MER's communal operations, including the PI leadership, resemble modern expeditions of field scientists, but the organization is much larger and distributed. The team's consensus approach is enabled by visualization, planning/commanding, and communication tools that promote a virtual experience of working together on Mars. Further lessons can be gleamed by comparing Apollo's organization and operations and then considering more broadly how technological advances relate to scientific projects.

Comparing MER with Apollo

The Apollo lunar surface missions share some aspects of each category of scientific practice, too. In contrast with MER, the Apollo traverses were much shorter and all activities were planned on a detailed schedule (at the level of minutes) months and even years in advance. Like MER, the astronauts were engaged in field science and deployed instruments that transmitted data for analysis on Earth. Although the astronauts were actually present on the planetary surface being explored, the analysis of samples only occurred when they returned home, so the findings could not inform their exploration—precisely the inverse of the MER situation, in which all instrument measurements occurred during the scientific exploration and no samples were returned.

From a scientific—as opposed to political—perspective, Apollo was a form of big science that replaced almost total ignorance (we wondered whether Surveyor would disappear in dust) by geological surveys and a theory of the moon's origin. However, deploying instruments and systematically gathering and documenting samples during Apollo were not preordained. With the engineering focus on getting people to the moon and the risks involved, the first conceptions of a lunar landing argued against scientific activities: "Somewhere, somehow, amid the 6 million pounds and 363 feet [of the Saturn V], we'd have to squeeze in a scientific payload."[9] Yet the astronauts' pioneering work was a credible beginning for lunar field science.

The assemblage around 1964 of field science specialists in geology, geochemistry, petrography, mineralogy, and geophysics to formulate lunar science plans is analogous to MER's science theme groups. Apollo astronaut training and mission simulation in lunar analog settings (such as volcanic flows on the islands of Hawaii), parallels MER's ORTs with a rover in a desert setting, showing that this practice of preparing for fieldwork is well established. In some respects, Apollo's "science backroom" (actually three science support rooms in the later missions) resembled MER's science assessment room for large-group planning and advising of EVA operations. Of course, on Apollo, scientists were guiding well-rehearsed human activities rather than programming "Observations" for a robotic laboratory. With the advantage of the almost negligible time delay, the backroom could participate in the ongoing lunar exploration in real time (with CapCom serving as their proxy). The experience of curating and sharing lunar samples in the space science community over subsequent decades parallels the concern in MER for gathering data for future generations, as well as the public and formal efforts to share and publicize data products and findings.

In summary, the Apollo experience effectively established a "total systems" approach to science operations for surface explorations that was further developed by decades of experience with planetary orbiters and flyby spacecraft. This operations background was reconceived for MER into an exploration system configured for virtual presence on an overland journey.

Relation of Technology to Scientific Advance

The nature and role of technology is an overarching theme that partly differentiates the categories of scientific practices (table 10.1). As Dyson has essayed, the progress of science is both "tool-driven" and "idea-driven," a history of tools and a history of ideas.[10] What we want to investigate and how we investigate shape each other. For example, the history of particle physics was first dominated by the use of optical detectors and images, and after 1980 "electronic detectors and logic" came to the fore.[11] Similarly, Mars exploration until the Viking landers was optical (i.e., direct-light photographs), but since then optical and electronic imaging (e.g., spectral scans) have been used in synergy, both in orbit and on the surface. Consider, for example, how the spectral analyses enabled by Odyssey and the improved optical images of the Mars Reconnaissance Orbiter guided the design and targeting of the Phoenix lander. MER demonstrates the synergy of Pancam (optical) and Mini-TES (electronic) detectors.

Placing such instruments on rovers makes a huge qualitative difference. The improvement from Pathfinder to MER in the combination of instruments, telerobotics, and scientific organization and process demonstrates how tools have changed the investigation of Mars, making it a field science. Putting Athena on a rover was driven by a scientific surveying method and an interest, namely that scientists carrying appropriate cameras and laboratory instruments could "follow the water," traversing on Mars to find where life could have existed. Now they had a virtual body that could not only sense and manipulate but also move and climb.

The relation of scientific tools and ideas is logically "both-and": neither dominates or controls the other. Rather, each provides context for the other to develop and flourish; as expressed by Cabrol, "The history of science shows that knowledge on any scientific question is shaped by the means of exploration and those means are molded by what we think the world is."[12] Scientific tools enable discovering new phenomena and formulating new theories, leading to new interests and then developing new tools to learn more. Accordingly, a major argument for developing the MSL is that a different suite of tools could better approach "what a person could do on Mars." Thus, the scientists' imagination of how they would conduct field science if they were present—what tools they would use (e.g., a drill), where they would go, and how they would analyze materials—drives the design of new technology. For example, Squyres said, "To test the hypothesis that there are organic molecules on Mars … MSL is going to drill centimeters into rock, extract powder, bring it into the vehicle, and search for organics with extraordinary sensitivity."[13]

With new tools, scientific communities sometimes experience a paradigm shift in theories and methods.[14] For example, the discovery of dynamic processes on the moons of Jupiter and Saturn caused by tidal forces has produced a kind of paradigm shift in the search for life—the energy required may be available without proximity to a star. The MER exploration system has enabled a new kind of field science, but this has not been a Kuhnian shift in how geologists, geochemists, or meteorologists understand the structures and processes of a planetary object. Rather, MER has changed how familiar scientific work can be accomplished, performing in situ laboratory analyses during a multiyear traverse while coordinating some observations with orbiting instruments. This combination of spacecraft, telerobotics, and instrument miniaturization is relatively new and just begins to suggest what we might accomplish with a broader system of automated, networked technologies.

In subsequent sections, I investigate MER's methodological shift further: is the virtual, communal, public nature of doing field science tenable and effective in its present form? What additional technology and perhaps different operational organizations should be considered? In imaging how actually being on Mars would be better, are the MER scientists merely expressing their frustration with a pioneering telerobotic system or are they beginning to articulate aspects of scientific inquiry—ways of manipulating, looking, moving—that could be done better by people working on Mars?

The Developing Practice of Mars Exploration: A Dynamic between the Self and Sociotechnological Context

In examining the scientific practice of MER and how the fieldwork is actually accomplished, we have viewed MER as a system in which technology on Mars and on Earth, roles, schedules, and even facilities synergistically relate. Examining how this system evolved during the mission, we can see how a sociotechnological context folds onto itself, as each aspect—personal relationships, procedures, tools—shapes the mission and is changed by it.

The Individual-Social Dynamic

To illustrate the relation between the scientists and the mission environment, let's start from a personal perspective. Personal interests and capabilities develop within the motivations and activities of the group in which individuals engage; in turn, individual contributions change how the mission is carried out. For example, Sims's ideas about software autonomy and virtual reality could be better articulated after the MER team's experience in controlling the rovers. Besides explicating this theoretical lesson, Sims opportunistically developed new tools to automate parts of the uplink process, illustrating his theories in context. Similarly, reflections from repeated experience motivated and justified Trebi-Ollennu's experiments with his colleagues in using new navigation software onboard the rover. As a researcher at CMU, he might have once just built what he liked, but in the context of the mission, particular ideas and capabilities are *selected* and become acceptable and valued by the team. Regarding targets and pace on the way to the Columbia Hills, Rice represented the surveying geologist's point of view, contributing to Des Marais's compromise innovation of the four-sol quartet. These are just a few illustrative examples of how MER exploration developed as a sociotechnical system through individual contributions from the science and engineering teams over the years during the mission itself.

The relation of the individual to the historical context is similar. Consider, for example, personal experience while the rovers were landing, events that could not be observed directly but only inferred from the spacecrafts' time-delayed reports. The reported emotional feelings, possibly common, flow from awareness of the importance of the mission within the longstanding historical effort to reach Mars and, perhaps more specifically, the significance to the entire Mars program of simply landing safely, given memories of past disasters. The landing occurred after years of difficult and sometimes frustrating work (e.g., getting the air bags to work); the successful ending provided an overwhelmingly powerful release. At the memorial ceremony, Charles Elachi, director of JPL, summed up the landing: "Spirit lifted the spirit of NASA when it really mattered.... That [landing] created such excitement, and you believed in NASA and what NASA can achieve, the things that are always 'impossible.'"[15] In such moments, the notion of a "mission" becomes evident not just as a name for an institutional project, but as a form of lifework to which individuals are intensely devoted. As Yingst said, working on a mission can feel like responding to a "calling ... in the same way some people are called to be a preacher." At the loss of a mission, one loses part of oneself, one's purpose; Yingst reported after the loss of Mars Polar Lander in 1999 that she felt devastated.

During the mission, everyone's conception of "what we are doing now" continues to be influenced by a historical context that combines personal, public, and mission-specific concerns. Subsequently, on retelling the landing story later, the moment has become a part of history itself; the event takes on further meaning as part of its place in the broader adventure—"arrival in the new land" and the start of the voyage of scientific discovery. The story of bouncing into Eagle crater becomes meaningful and worth telling because of what was found there, but it is as full of emotion as if the teller had personally struck a hole-in-one.

Notably, in retelling the landing story, people emphasize the social setting in which their experience occurred, ranging from a suddenly overflowing cruise mission support area at JPL to the stark privacy of a strange apartment in Pasadena.[16] This memory of "where I was, what I was doing" shows further how individual experience (and hence memory) is conceptually bound to broader social contexts. Indeed, Squyres described how they exploited such temporal and historical associations in naming places on Mars: "Bastille Day we gave everything French names. Thanksgiving it's cranberry and drumstick and wishbone … *(laughs)* Rock 1750A—nobody remembers what that was. But you show me the rock, Cherrybomb. I go, 'Cherrybomb, fireworks. Fourth of July … okay, it was that panorama.' We have used that trick to *great* effect."[17] They have transformed a Mars traverse into a chronological "memory walk."

The relation of individuals to the social setting is also evident in the treatment of MER photographs as art. On the one hand, looking at an image involves a personal perceptual-affective experience. On the other hand, the image's meaning is broadened by social activities in which photographs are shared, such as museum exhibits, publications, and presentations to friends. The response of other people then affects the scientists' own appreciation and sense of accomplishment.

Arguably, this dynamic among the individual, the group, and technology has always been integral to the experience of exploration and adventure in a wide range of scientific projects (table 10.1). New technologies enable new types of measurements, but also enable people to act more effectively as a group, with what Squyres called "the collective brain." For example, at the most mundane level, the MER photos are made available through a sociotechnical system with interleaving scientific and engineering teamwork, using technologies—image processing software, communication, and Internet access and search tools—that didn't exist a few decades ago. Within the mission, opportunities to see the images are provided by and occur within the activity of the collective, shaped by a complex of organized processes, facilities, and technologies (e.g., figures 5.5, 6.7, 7.2, 8.3). Future missions might learn from MER to deliberately design the work system to enhance collaboration, focusing especially on synergy of the instruments and visualization-planning tools like SAP and Viz that promote and support conversations across the disciplines.

MER shows us conclusively that a phrase from the popular *Star Trek* television show—"where no man has gone before"—has another literal truth: exploration of most extreme environments cannot be done alone. MER lies along the path of a long history of using technology to extend the group's reach. For example, the navigation and shelter capabilities of the sailing ships of the eighteenth century enabled long-duration, communal journeys. The risks of circumplanetary travel remain so extreme that centuries later, individuals still make the news by attempting passage alone.

In summary, the MER voyages required people to work together for a long time, making salient perhaps more than most previous missions the individual experience of participating in a large, unified team. Instead of perhaps a smaller story of being part of an instrument group and that group's struggle within the mission, the MER scientists relate a much broader, often historical perspective of what it means to be exploring a planet, "gathering data for the ages" in a manner everyone could share.

Aligning Thoughts, Tools, and Actions with the Rover

In considering how individuals relate to the mission, their obligations are most insistently defined by the rover's current state and setting. Ideas for using the instruments necessarily must orient to the terrain (e.g., avoiding sand traps or sliding down crater slopes), the health of the rover (particularly the power available), and the opportunities these offer in combination with the instruments for advancing the scientific investigation.

Although everyone appears to readily adapt mentally to the embodiment of "being the rover," the imagination of "where we are" or "what we could do" must eventually be formally registered on a surface coordinate system (using a form of virtual reality in tools such as Viz and SAP) that is precise enough to productively control and predict rover behaviors. Combining personal imagination of what you want the rover to do with the tedium of programming action on Mars leads to a stark realization of now narrowly the rover's capabilities squeeze the mental and physical experience of exploring the surface: one rover planner referred to the experience of driving on Mars as "trying to make our way through a dark cluttered room with nothing but a flashbulb."[18] Those of us who were not working on Mars see the planet through the photographs (e.g., figure 9.1), but the scientists are also aware of the blank spaces in between that have not been imaged or studied in such detail.

Accordingly, the team's practice—how they interact and use tools, as well as their ways of talking—should be understood as solving essential problems: where are we on Mars? Where are we in our investigation? How are our tools performing? What should we do next? At first, the anthropomorphism of rover behaviors (e.g., referring to a "siesta"), personal projection into the body of the rover, and metaphors such as "robotic geologist" might seem humorous or even nonscientific. When I first advocated "clear speaking about machines,"[19] I argued that such metaphors could distort robotic research as well as the public's understanding of current technology.

The function of anthropomorphic metaphors for the scientists in designing and operating rovers like MER is a different matter. An insightful analysis followed by considering the sociotechnical and cognitive context in which different perspectives play useful roles (chapter 6). In particular, "becoming the rover" (first-person view) helps the group construct a shared conception of "what are we doing now"—their exploration activity as one body. This shared perspective provides agency both for individuals in coordinating perceptual-motor behavior, as well as for the group in cooperatively using the instruments and resources. Deliberate anthropometric design of the instrument arm, cameras, and mobility made the first-person perspective possible and is exploited by the visualization tools. The second-person perspective, constituting a partnership or "symbiosis," recognizes what has become on MER an almost matter-of-fact automation of routine tasks, such as operating an instrument and navigating to a target. The third-person perspective is a narrative shorthand ("Spirit investigated") that personifies the rover, publicly objectifying the group's accomplishments and reinforcing focus on the team's investigation of the planet, which transcends and integrates individual interests. Recognizing the value of these distinct perspectives helps us to understand how the MER *exploration system* of people and tools accomplishes the work as an integrated whole.

The success of MER's design is measured by the quality of the field investigation, not just the list of findings, but the extent of the surface area surveyed and range of features that were examined and analyzed (the time required and cost are independent metrics). This scientific result was possible because the design of the MER exploration system fits how people think and work individually and collectively. The first-person plural way of talking ("are we going to go or are we going to stay here?") shows that the visualization tools are aligning personal mental and social perspectives on the rover's state and future actions (figures 6.5 and 8.7). Through gestures, tools for labeling targets, and formulating plans (figure 7.3), individuals represent their interpretations and intentions such that each person is aligned with the rover and the scientists and engineers are aligned with each other.

In many aspects of the mission, the "one instrument, one team" concept looms in importance: in effect, the Athena hardware, as configured on the rover, is a *collaboration tool*. MER gives new meaning to this phrase, referring not just to a means for exchanging information and facilitating meetings (e.g., a shared display "hyperwall") but to a system for integrating multidisciplinary perceptions and actions into a joint activity on another planet. The channeling of all the individual ideas and desires into one sequence of rover actions fosters the image of a single entity acting on Mars, a surrogate who carries out the group's plan and thus comes to represent them in the third-person account on the JPL website.

In considering what accounts for the quality of the fieldwork, I concluded that the central capability is enabling the scientists to be *agents on Mars*—effectively coordinating a sequence of actions—looking, manipulating materials, analyzing spectrally, and moving. Agency on a remote surface is supported especially through tools for visualization and programming robotic behaviors (observations/activities). Combined with human imagination—relating images, bodily projection, and conceptions of action—these tools enable the scientists to experience *virtual presence*. Finally, this experience of being an explorer on Mars is reinforced and sustained through *daily science commanding*, the ability of the science team to receive data and reprogram the rover every sol. Daily commanding was first justified by efficiency and urgency, but in practice kept the scientists' engaged and promoted science-engineering collaboration in what was otherwise, compared to fieldwork on Earth, a tediously slow advance through the landscape. This relatively complex relation among the MER exploration system's objectives, features, and methods is summarized in figure 4.3.

The advance in personal computing since Viking is obvious to even nonscientists, but few may realize how the advance in mission operations tools over the past decade alone has dramatically affected our ability to operate a remote laboratory. Although some of MER's sequencing tools and processes were prototyped and demonstrated on Pathfinder, the advance since the late 1990s in web-based servers and browsers, databases, visualization graphics, and programming languages has enabled much more rapid systems integration for practical purposes. In particular, MAPGEN and the Constraint Editor, two of the sequence generation tools used by engineers on MER, were developed further to enable Phoenix mission scientists to use these tools collaboratively in their planning meetings.[20] Such tools move engineering constraints previously relegated to the SOWG and sequencing process (e.g., adjusting the priorities based on summative power, timings, memory, and bandwidth constraints) to earlier in the science planning process. Different interests and instrument operations can be scheduled together, as opportunities and conflicts are visibly obvious in the group science meetings. Accordingly, discussion of plans with remote members and engineers enables collaboratively configuring more instruments and operations and takes less time while requiring fewer manual steps and people in the process. Similarly, adapting SAP for distributed operations increased the number of scientists able to actively contribute after the initial ninety sols by bringing the tools directly to institutional groups.[21]

As a result of MER, the engineering practices within JPL have changed, how scientists work on a NASA mission has changed, and how we conceive of people and rovers working together has changed. So what can we learn more generally about planetary exploration from the MER experience? How does MER inform future strategies for exploring Mars?

Organizing People and Configuring Robots

A perennial favorite question for conference panels and space program editorials is, "What are the best roles for people and robots in surface exploration?" The surface of Mars is of course the most likely area of contention, but asteroids, Phobos, and the possibility of mining on the moon are of interest, too. Given years of unprecedented experience in scientifically exploring a planetary surface using the Mars Exploration Rovers, can we generalize any principles for future investigations? In this section, I consider organizational lessons, the relation of people to rovers, and the strategic implications for exploring an entire planet.

Organizational Implications of Surface Missions

MER has demonstrated a sustainable method for operating rovers for years with a distributed science team. The duration of the Pathfinder mission was comparatively brief and the investigation was confined to the landing site. The MER mission constitutes the first time we can draw conclusions about telerobotic operations in multiyear missions roving through large regions (on the order of at least tens of kilometers from the landing location). Engineering lessons aside (which go beyond the scope of this book) we can conclude that Athena's design has been resoundingly confirmed: the scientific work is fruitfully organized around the metaphor of a "roving scientific laboratory"—a coherent set of programmable instruments that provide ways to see, clean, and analyze materials on a mobile platform that enables traveling to and over features (and touching specific targets) whose morphology (size, shape, solidity, and orientation) cannot be fully anticipated.

To put this conclusion a different way, compare a rover to an orbital or flyby mission with remote sensing in which instruments are used opportunistically or by predetermined program. For in situ robotics, the instruments move to different locations and the team thus moves together, regardless of whether individuals want to be in a certain spot. With "two boots on the ground," the field scientist must be able to see a certain distance (Pancam), resolve fine details (MI), navigate around obstacles (Navcam), analyze chemistry close up (APXS, Mini-TES), and break or scratch rocks (RAT). Contrast these to Galileo's instruments for detecting the composition and properties of gases, particles and fields, lightning and radio emissions, and so on. As Squyres emphasized, a rover's instruments must work together, because the field science requires repeated surveying for interesting features, zeroing in opportunistically on targets, and then analyzing them closely. The broad topographic features provide contextual information (e.g., the distribution of iron or sulfates, which can be detected even from orbit), though most of the mission's questions about water require microanalysis of an outcrop, abraded surface, or soil patch. When instruments are developed by different organizations, as is common for robotic missions and has occurred on MER, the integration of the instruments requires ongoing collaboration during design and testing, with a coherent vision of how the whole suite will be operated.

The scientific exploration of a surface not only involves complementary instruments but also requires retargeting them opportunistically (figure 4.3). Compared to orbiters and flyby missions, rovers enable a qualitatively different kind of exploration in which what is worth exploring and especially how much time is devoted are determined from day to day during the investigation. Although spacecraft such as Cassini enable exploring a region of space and we make discoveries that influence future targets and trajectories (e.g., focusing instruments on the dark areas on Titan and the geysers of Enceladus), the possible orbital routes are both limited and periodic—plus, of course, a spacecraft can't be stopped in one place or its instruments redirected to centimeter-size targets. Missions to asteroids and comets can involve multiple destinations, but schedules are determined years in advance.[22] In contrast, a programmable mobile laboratory provides daily flexibility for the scientists to adapt the mission to what they are learning. Consequently, the tools and people involved in retrieving, distributing, and analyzing the data must feed back fairly tightly into the planning and reprogramming process.

Turning now specifically to the team's organization, the MER scientists have demonstrated how to carry on the constrained work of exploration and detailed investigations through science theme and long-term planning groups, interacting in an originally daily process of science context, science assessment, SOWG, and end-of-sol meetings. Arguably, the coherence of this process was enabled by having a single PI and rotating sequence of SOWG chairs (all senior scientists) who oversaw the planning and sequencing process with the engineers. Matijevic concluded, "This experience is a great model for how we should be doing things from this point on." With results that went well beyond the requirements of the nominal mission, MER's leadership, organization, instruments, and software facilitated the consensus and endurance required for a long-duration, remotely controlled expedition.

Although all planetary missions require cooperation between scientists and engineers, the examples given by Matijevic and Trebi-Ollennu show that these interactions on MER have gone further to become real collaborations. Together, they negotiate methods within a discussion of proposed operations that reveals the scientific intent and longer-term objectives. We must remember that the scientists and engineers are indeed different organizations, with people employed by different institutions, having different technical perspectives. For example, Trebi-Ollennu's comment that "the SOWG Chair would try to touch base with all the instruments" before the engineers sketched the rover sequence shows how the SOWG chair (a scientist) successfully communicated with the individual engineering groups, whose language is that of instruments, not cross-bedding, jarosite, or spherules.

Collaborative investigations with engineers and scientists is of course not unique to space exploration (e.g., consider research and development in oil exploration). Such collaborations have occurred throughout the space program. For example, in managing Hubble, Space Science Institute "staff included scientists who specialized in calibrating and understanding the Telescope's instruments, and engineers who specialized in merging thousands of approved observation requests into an efficient observing schedule."[23]

Yet each technology provides its own opportunities for learning new operations concepts. For example, the Phoenix lander, planned for only a single Mars polar summer lifetime with key laboratory instruments capable of processing only a small, fixed number of samples, posed different engineering challenges than MER.[24]

Matijevic said that MER experience has given the engineers "a basic understanding of how you do surface science," namely, how instruments were logically sequenced in seeing and moving to a feature, then examining and analyzing targets. This grasp of the pattern and rhythm of fieldwork has helped managers and engineers plan MSL operations to further enhance collaboration with the scientists. MSL could provide an important counterpoint to the management of MER's design and operations, by a combined project scientist–PI role and having individual instrument PIs, from which further generalizations might be drawn about how rover instruments are designed and surface operations are managed.

The Relation of People and Robots in Exploration

Given the costs of developing an infrastructure for long-duration human spaceflight to Mars and the risk involved, exploring a planetary surface through some combination of telescience and "autonomy" like what MER provides is appealing. Obviously, the rover methodology succeeds in providing relevant data and enabling flexibility for scientific exploration without any physical risk for people. This combination of cost, return, and safety suggests that we will want to continue to improve and exploit rover technologies for the indefinite future for the rocky planets and moons.

But with ongoing efforts to develop sustained human exploration on the moon and beyond, the future relation of people and robots is unclear. Panels are convened at workshops with questions like, "When does the human become the tool of choice for solar system exploration?" and "How should the ratio of humans to robots change over time to meet that goal?"[25] From both a scientific and an engineering perspective, the question most often posed in the first decade of the twenty-first century has been to determine "the right mix" of people and robots. It behooves us therefore to learn as much as we can from the MER experience.

To start, the two example questions given here—which are not fabricated, but were actually raised by senior computer scientists—are technically and philosophically confused. Consider in particular the choice these questions offer: "human tools" versus "robotic geologists." Notice that the roles of people and machines have been switched: now the geologist is a tool and the robot is a geologist. Equating people with machines by putting them in the same category is a well-known example of reductionism. That is, it simplifies the nature of people, equating the brain with a digital computer and human knowledge with computer models. Such a conceptual gloss was perhaps acceptable in early stages of psychology—for example, in early twentieth-century analogies equating the brain with a telephone switchboard. However, neuropsychology and cognitive science have moved far from this position, showing that although you might call the brain a "machine" because it is composed of physical "parts," it is unlike any machine we can currently build. For example, human memory is not like stored things, but more like processes that can be reactivated and recomposed in ways related to how they have operated together before. That is, the "parts" have an organization that is dynamic and always adapted. The mid-twentieth-century theory that human knowledge is stored in memory exactly like a computer program is technically incorrect.[26]

Given a question referring to a "human tool," our first response must be to correct the misconception that a person could be used as a tool. We send people into space not to be servants or to do the work of machines but, first, to live and thrive in space and, second, to do what machines cannot do. We must be clear about the difference between a person and a tool: a person perceives the spectra measured by an APXS instrument or the Mini-TES on MER. But a person cannot hold the APXS still in a ten-hour "stare," as its application requires, in the manner of the IDD arm. The list of such contrasts between human and instrument-related capabilities is quite long. In carrying out operations a person could never do, MER is more than a physical surrogate of a geologist. Cognitively, no strategic, interpretive work can be delegated to MER. As discussed, the comparisons of "what a person could do" are incomplete and somewhat obscuring because they focus on a subset of MER's operations such as rolling over a desert plain, breaking open rocks, and climbing hills.

When we study the MER mission and how people actually work with the rovers, we find that the looming exploration problem is not how to reduce the number of scientists from 150 people managing two rovers—instead, it seems that a better conclusion is that organizing such a "collective brain" is a major accomplishment. This success suggests that we reach further: how can we keep the larger scientific community productively involved? That is, can we *increase* the number of scientists using one rover? And the efficiency problem is not how many rovers could a single engineer control but first, much more practically, how do we reduce the 75:1 ratio of engineers for each rover? The cost-benefit trade-offs are so stark that how to invest in technology is not obvious—a new design might save months of surface operations time and perhaps millions of dollars in engineering operations salaries but result in losing a multibillion-dollar machine. The loss of Mars Polar Lander suggested that design decisions (e.g., omitting descent telemetry) also need to be made in the context of broader mission programs, investing for incremental improvements, and learning from mission to mission.[27]

The advantages for scientific exploration of MER's commanding turnaround over Viking are obvious (table 2.1 and figure 4.1), but improving automation further requires a closer look at the scientific work. Considering the different ways in which people can use robotic tools, including being present at the same time and place (figure 6.9), the question then becomes a matter of objectives and constraints. What are the scientific requirements for gathering data? What is the extent of the land area and its topography? What are the budgets and time schedules within which the mission must be configured and operated? For example, we might list surface work tasks as including survey and mapping, prospecting, and searching for life. Each of these requires investigation of hypotheses, choice and application of instruments (e.g., is a mini-Mini-TES stare of four minutes sufficient?), comparative analysis (e.g., with known minerals), perhaps curating samples, and documenting the data and interpretations.

The practical issue at hand is what kinds of robotic tools will be possible in the next few decades and how we shall operate them. The question "How should the ratio of humans to robots change in the future?" is as meaningless as "How should the ratio of cranes, backhoes, and dump trucks to workers at building construction sites change in the future?" What is the nature of the work we are trying to do, the money available, and the time allowed? A construction site is covered by people who are guiding, controlling, and holding tools and heavy machinery of all kinds. If we seek to replace construction workers— say, for preparing sites on the lunar surface—we will need something like a robotic backhoe or truck, which might be operated from Earth. Related proposals involve investigating Mars from a base on Phobos. The variety of options is perplexing.

Imagine the early sixteenth-century Europeans, if they had all of our present-day knowledge of machinery and remote operations and, informed by fine satellite photos, were discussing how to settle North America—it might feel as discombobulated and bewildering as our early twenty-first-century disorientation about destinations and rocket design.

It seems clear that cost is the dominant issue, not risk to people. If robots were truly so capable, and cheaper and safer than sending people to Mars, why aren't we using them in coal mines, rather than endangering human lives? Surely it would cost less to develop a robotic miner than to send an actual robotic geologist to Mars. One reason we do not replace human miners with robots is because of economics and our values: the loss of human lives is tolerated more than the expense of robotic miners. As one commentator put it, telerobotic miners are possible but financially irrelevant: "But why would a company spend millions developing or buying a robot that doesn't replace a miner or save them any money? The robot still has to be teleoperated by someone. Mining companies already have mining equipment, which would be cheaper to use than letting the miners have the luxury to telecommute."[28] We could defer mining until we have effective robots, but we choose to extract the resources now. We also have many people around the world willing to risk their lives to earn a livelihood as miners.

In short, arguments against human spaceflight because "it's too dangerous" are specious—nobody's forcing mining companies to use robots, yet more miners die every year than in the history of the space program. In the United States, an average of thirty-three miners died each year from 2000 to 2007. In October 2006 alone, China reported 345 mining fatalities.[29] If spaceflight followed the logic of mining, few people could insist that we should send only robots to Mars: people are willing to risk their lives on Mars, and robotic geologists do not exist. Nevertheless, nobody wants to lose a multibillion-dollar investment, compounded by the delays, lost jobs, and redesigns that a failure investigation typically requires. Simply put, even with compelling reasons to work on Mars, we cannot afford to send people until we can be relatively sure that they will stay alive and be productive.

For some scientific exploration objectives, robotic spacecraft and rovers cost considerably less than human space flight, useful technologies are already available, and the methods are effective. The issue for Mars science then becomes how we improve and combine technologies to make a yet more effective scientific exploration system, given the challenge of exploring an entire planet.

Exploring an Entire Planet: Staffing, Mobility, and Strategy

For the first time in the history of the space program, we have long-term experience working with a programmed robotic laboratory to scientifically explore a planetary surface. Can we extrapolate from the science team's operations, the mobility of the rovers, and the exploration strategy to draw implications about the future exploration of Mars?

How Many Engineers are Required?

One place to begin is the staffing of the MER mission. JPL's workforce, mostly engineers but including management and a few scientists, was about 275 people at launch, averaged 200 in 2004 the first year of surface operations, fewer than 100 in 2005, 85 in 2006, and leveled at approximately 60 from 2007 through 2009. After the loss of contact with Spirit, about 50 people were working at JPL on MER.[30] These workforce reductions were mostly required by reduced budgets; the effect was to reduce the science data returned in a given period of time.

In particular, operations were significantly modified by shifting from living and working on Mars time after the first ninety sols (complicating the single-sol feedback cycle of commanding and analyzing results) and reducing operations to five days a week (slowing the pace of the work).[31] The engineering effort was reduced from fourteen planning days per week (two rovers times seven sols) to an average of eight planning days per week (both rovers).[32] This schedule resulted in fewer sols for driving or using the IDD, because the out-of-phase Mars-Earth operations require planning when the current relevant state of the vehicle isn't known or hasn't been achieved yet; generating multisol plans for weekends involves a similar problem.[33] Engineering time was also reduced by scheduling housekeeping and science activities serially, rather than the initial parallel operations that required careful checking—again at the cost of accomplishing less in a single sol.

The shift to Earth time was in part facilitated by the use of "glueware" to connect data and command processing steps. For example, Sims's tool for automating routine operations of instrument leads for uplinking commands to MER enabled engineers to potentially split their time on more missions.[34] Experience doing the work during the prime mission also enabled engineering tasks and roles to be refined and libraries of command sequences to be collected and reused.[35] The improved planning tools, processes, and knowledge enabled generating commands for up to three sols in a single day.[36] The rover can drive only once in such multisol plans, but the science instruments can be used during the other sols.[37]

A statistical analysis by Trebi-Ollennu and Diaz-Calderon of rover command sequences for 1,000 sols showed that "there has not been any significant reduction in complexity of a sol's plan as result of moving to Earth-time planning." Furthermore, the size and membership of the rover planning team were relatively stable. Consequently, the reduction in science return between the prime and extended missions is attributable to the loss in a perfect daily commanding schedule originally enabled by synchronization with Mars time and working seven days a week—not by loss of staff to do the actual planning. Hardware failures were deemed to have minimal effect but required more complex IDD plans, which the team's experience enabled constructing in less time, with a pronounced increase in efficiency after the first year.[38]

In total, the efficiency improvements from automation and learning were essential for coping with staff reductions forced by reduced budgets, but the result was still less return of data from Mars over a multiple-sol period during the extended mission than had been possible in the first ninety sols. The budget cut aside, the stress of the daily operations, living on Mars time, and separation of the scientists from their homes probably made the shift to Earth time and consequent slower pace inevitable.

The JPL workforce data suggest that a core team of about sixty engineers may be sufficient for operating two rovers like MER on a five-day-a-week, Earth-time schedule. During this same period, the reduced budget and slower pace resulted in reducing the number of scientists involved to about 100 in 2011.[39] Considering the automation and team's experience, how many of the original more than 200 engineers and 150 scientists would be required for a seven-day, Mars-time schedule today? This number is difficult to predict without putting the pieces together and knowing who would be on the team. MSL operations will provide some data, but comparison with MER will be complicated by the substantially different instruments and power system enabling twenty-four-hour operations. However, we can reliably expect automation and learning will compensate in part for budget reductions during the course of the extended mission.

How Much Time Is Required to Explore an Area?

From the scientists' perspective, the most salient issue in MER operations is the slow pace of the work. In examining the scientists' experience of frustration, we found that pace is relative to personal expectation from field experience and scientific needs. Expectations probably need to adapt further in doing fieldwork remotely. But we can also approach the work on Mars differently, such as decoupling spectral analysis and surveying by using multiple rovers. With improved planning and programming tools, and given sufficient communication relay opportunities, smaller groups living on Mars time could receive data and replan operations during the martian day, allowing multiple reconnaissance surveys in different directions at a much faster pace. Here I address in particular whether a different rover design, with more power and better mobility, would have been advantageous for the investigation of Home Plate.

The story of Home Plate reveals the difficulty of relating results and schedule to the rover's capability; so much depends on what's there to be discovered, the climate, and the terrain. Reflecting on the duration of the campaign twenty-two months after Spirit arrived in the Home Plate area (about 90 m diameter; figure 10.1/plate 25), Squyres said in 2007 that with a more capable rover, "We would have left Home Plate a long time ago." On the other hand, "It's the most interesting place that Spirit has been the entire mission. There have been some *astounding* discoveries such as Silica Valley; it took us 1,200 sols to find that."

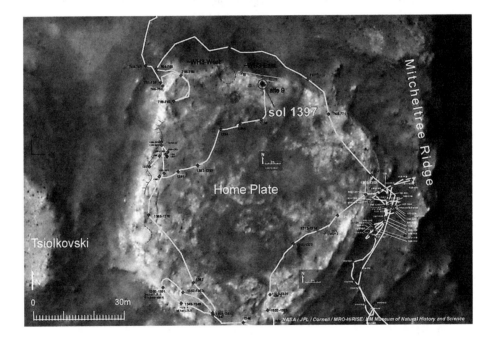

FIGURE 10.1/PLATE 25
Traverse map of Spirit in the Home Plate region behind the Columbia Hills, drawn on a magnified HiRISE image: as of late 2007, about fifty scientists using one rover with the help about sixty engineers had explored a region of about 6000 m^2 for more than eighteen months. *Image credit:* NASA/JPL-Caltech/UA/Cornell/NM Museum of Natural History and Science. *Source:* http://photojournal.jpl.nasa.gov/catalog/PIA10129 (accessed December 15, 2007).

CHAPTER 10

Speaking in late February 2006, a few weeks after Spirit arrived at Home Plate, MER project scientist Joy Crisp said that they wanted to stay only about five weeks, "We want to leave Home Plate no later than sol 780 (about two weeks from today)." The team was hoping to arrive at McCool Hill about mid-April.[40] So when the team first arrived at Home Plate, they really had no idea how long they would be there. Crisp said five weeks, but it was over three Earth years and ultimately became the rover's resting place.

According to Sims, the team was always planning to move on, but they continuously made new discoveries worth investigating:

> Our history at Home Plate reminds me of an old country song line, "Thank God for unanswered prayers." If we hadn't had wheel problems, we would have probably moved on, and my guess is we would have not gotten back to Home Plate. I view the Home Plate experience, and especially Silica Valley, as the most incredible experiment in scientific discovery that we have had in MER. Each sol we were expecting to depart Silica Valley in "about two weeks," and that lasted for a year of new finds and compelling new reasons for staying just a bit longer.[41]

Eventually, it became necessary to park Spirit for the winter, preferably on at least a 25-degree slope, which was best handled by staying in the area. Ironically, if the team hadn't spent so many weeks studying the silica, Squyres said, "We would have made it down to von Braun or Goddard or something like that for the winter, and frankly, probably would have died there because the slopes wouldn't be steep enough." Spirit eventually died at Home Plate for the same reason (getting stuck in an unfavorable location), but at this point in the mission, with continuing surprises and their strategy of being systematic, the consensus was to stay.

The simple facts—an area 90 m in diameter, which is more than 6,000 square meters or somewhat larger than an American football field, explored by more than a hundred scientists and engineers using one rover for over two years—shout back for consideration: what if we had a rover with MER's instruments whose operations weren't so strongly limited by power? How much time should it take to explore such a region? The scientists interviewed found this situation so strikingly different from the reality of using Spirit that most said the question is too hypothetical to be answered meaningfully. Squyres summarized this perspective in March 2008, when the productive work was still under way:

> Our operational decisions are made on a continuous basis, on the basis of what we know at the time. With six good wheels and no power issues, our traverse path at Home Plate surely would have been different from what it was. But none of us can guess what it would have been, because none of us can guess what our knowledge would have been at the time we had to make critical decisions. Would we have even driven through Silica Valley if the rover had its full hill-climbing capability? I have no idea. And if all six wheels were working, would we have missed the serendipitous discovery of very high-silica soil deposits? There's no way to know.[42]

So one cannot strictly predict how the team's voyage on Spirit would have played out if the power were significantly different. But future rovers like MSL will have more power, and based on MER's longevity, long-term planners will surely look further ahead during a mission to project and argue for alternative investigations (indeed, a traverse plan for Gale crater was posted a year before MSL's planned landing). Spirit's limitations aside, suppose that a precise plan were required; how might a long-term planner approach the investigation of this area? Relying on "a combination of prior experiences and also the nature of Home Plate itself," as determined in those first few weeks, Des Marais speculates how the traverse might have gone with the Promised Land beckoning beyond:

> We did a reasonably thorough study of one small area on the northwest part of Home Plate. I assume that we might have done two more studies like that at two additional locations along the margin of Home Plate, for example, on the southwest and east sides. And we had also discussed driving along the more exhumed western margin of Home Plate, imaging it like we had imaged the bedrock in Eagle crater at Meridiani. I figure that ninety sols would have been required to execute these activities and complete the trip to the southeast edge of Home Plate, on our way to McCool Hill. But we wouldn't have observed the three-dimensional perspective of Home Plate's composition as completely as we actually did. More significantly, probably we would *not* have discovered the high-silica deposits, arguably the single most significant discovery of Spirit's mission.[43]

The difficulty of predicting how long they would stay at Home Plate when they first arrived and Des Marais's conclusion about the possibly missed discovery if they had followed even a ninety-sol plan reaffirm the uncertainty of surface mission planning and the serendipity of discovery. The challenges of the terrain relative to the rover will remain largely undeterminable until traverses are attempted. The combination of local knowledge and circumstance that affects both long- and short-term plans is unpredictable. Even with highly educated guesses, which are necessary before and throughout the mission, what we will discover and how long it will take can only be speculated upon. So regarding how much time is required to explore an area, Squyres sums up the lesson learned from MER: "There's no way to know." Looking at that central area in the region circumscribed by Spirit's traverses (figure 10.1), who can say what surprises remain?

Another dimension for speculation involves better reconnaissance and precision landing. Home Plate is far from the landing site in the flats of Gusev and the original objective of the mission to study an (hypothesized) ancient lakebed. Can we better target landing sites in the future by aerial or orbital reconnaissance? Trebi-Ollennu suggests these alternatives:

> Exploring Mars only using surface assets to do detailed study obviously is going to be cost prohibitive. Because you can see that if we had gotten good orbital images when we started, we would have landed closer to Home Plate, and we would have gone into this Silica Valley. Although I would prefer more robotic assets on the surface, I do think that a good range of orbital assets, and probably balloons or airships, in combination with surface robots would be an efficient way to explore the planet.

It might seem at first that a more powerful, larger rover like MSL is always better, but Trebi-Ollennu says, "For us to explore the area that we've gotten the most science out of for Spirit, we could have done it with a much smaller robot if we could have pinpointed the interest at Home Plate" and were able to control the landing precisely.

With an area as large as Mars, we need to know where to go. Our experience at both Gusev and Meridiani suggests to Trebi-Ollennu that after doing "a lot of traveling, where we find good science are very, very small places." The question is then whether we can identify these targets well from orbit and how far apart they are. This will then suggest an exploration strategy in terms of the size and number of rovers to deploy.

Orbiters, rovers, and landers have increasingly related roles. For example, high-resolution orbital imagery (from the Compact Reconnaissance Imaging Spectrometer for Mars, or CRISM) helped in targeting Opportunity; in particular, according to Arvidson, the detection of hydrated sulfate along the rim of Santa Maria was "the first time mineral detections from orbit are being used in tactical decisions about where to drive on Mars."[44] But orbital sensors need to be improved to identify the "small places" to sample. According to one report, the silica-rich area of Home Plate would not have been visible in MGS TES data because the area is too small. Steve Ruff, a research associate at Arizona State University overseeing the Mini-TES research on Spirit, said that no existing orbital spectral instrument could detect such materials: "It's very contaminated by soil. The sand that blows around and erodes this stuff is trapped in all the crevices and crannies of this rough-textured outcrop. But we can sort of see the meter-scale shape of these little ridges of outcrop in the HiRISE image of Gusev in the Home Plate area."[45] Interestingly, some of the key areas of scientific interest at Home Plate, which led to targeting MGS's TES, were discovered from the rover itself by a combination of Mini-TES scans from tens of meters and by imaging tracks created by its lame wheel.

Phoenix illustrates the benefit of a fixed lander in a homogeneous area ranging for hundreds of kilometers that had been selected from orbital studies of the icy terrain. Given the objective of confirming and characterizing the surface ice detected from orbit, the randomly determined reachable workspace at the landing site, on a plain of uncountable polygons, was probably as good as any other. The only reason to move might have been to avoid any worry of contamination from the descent rocket engine.

One way of thinking about the problem of exploring the entire planet of Mars is that finding new materials and formations is a giant egg hunt. As we look at Mars today, we can imagine that the eggs—surprising or useful mineral deposits, ores, geological formations, and the like—are scattered about. Our problem is that we don't know where they are. Going a step further in this analogy, there may be certain kinds of eggs to be found, for example in ancient lakebeds or areas with relatively recent water flows. We don't know what kinds of interesting places are present, how many there are, and how they are distributed on the surface. For example, how many places on Mars are like Home Plate? Without such knowledge, we can't know how many spots we need to visit before we can characterize the geologic and climatological evolution of the entire planet. Our problem is that designing an appropriate scientific exploration strategy—how we relate people and robots on missions over the coming decades—depends on the nature of the environment we are exploring, and that's what we are trying to find out! This is where the egg hunt comes in: by simulating different egg distributions in a computer program, we can evaluate and compare different methods for exploring Mars on a large scale. Nobody has carried out such a study yet; we are just learning enough to know to ask such questions and to think about how to answer them.

In reality, the rocks, soils, and deposits on Mars are not like "eggs," whose identity and value is immediately recognized. Rather, as Sims says, "When you find something on Mars, you only have a hint that it might be useful. It takes a lot of time and a lot of added pieces of information before any one piece of information makes sense." Building on the egg hunt idea, he suggests instead the metaphor of a jigsaw puzzle whose pieces are scattered about the terrain, making the work more like archeology: "But there are extra pieces lying around belonging to many puzzles. So finding one might weave together with the puzzle we are thinking about, or it might be a totally different puzzle and story entirely. Also, all the pieces are caked in mud. You have to first carefully clean off the mud before you can even see what it is."[46]

Sims's point can be elaborated upon by focusing on the analytic work of "removing the mud" from APXS data. In evaluating the quality of the science, rather than the timing, the pivotal issue is the inquiry that the technology permits. The technology need not support the actions that people would do if they were actually on Mars.[47] Here we need to help the scientists articulate their frustration into specific sensing, manipulating, moving, and data-filtering activities that might become functional designs for future robotic laboratories.

Designing Future Field Science Technology for Mars

As we have seen, when reflecting on their field experience on Mars, the scientists tend to compare MER to what people could do, such as abrading rocks, climbing, and moving to another location. But to design the next generation of mobile robotic laboratories, we need to focus on many more aspects of their work. We should start by analyzing what the scientists can't do *using MER* relative to their investigation of Mars or, put another way, what MER can't do (at all, well, or quickly enough) that would have made the field science more effective and/or efficient.

A proper evaluation of MER's limitations will surely occupy many reports and meetings, but we can anticipate and illustrate here some broad considerations. First, to ensure that the capabilities of the laboratory are effectively deployed within the region and time available, better *surveying* is required, that is, applying the field method described by Des Marais, which involves touring a site (perhaps from the air or orbit) before doing detailed analysis, rather than MER's linear no-going-back strategy. Better *physical coordination for precision sample handling* is required to deliver specific samples and quantities to instruments (e.g., apply APXS to only a single "blueberry" spherule; on Phoenix, precisely deliver a particular ice sample and amount to the Thermal and Evolved Gas Analyzer [TEGA], a combined oven and mass spectrometer). Better *mobility* is obviously required to avoid getting stuck in sand for weeks, to handle slipping (particularly on crater walls and to reach canyon floors), and to navigate to designated locations and apply instruments to targets with less human guidance. Better *scraping, breaking, and deep drilling* are required for exploratory sampling (e.g., Opportunity's RAT wore out; Phoenix couldn't effectively break up the ice in shallow trenches, and the scientists might have needed to go meters to show a hypothesized gradient[48]). *Curating samples* (carrying or caching identified soil, rocks, or fragments) would enable comparisons as interests change during the mission and as additional instruments are brought to Mars or for later returning samples to Earth. *Larger or heavier instruments* would permit different or better atomic and molecular analysis (a rationale for returning samples to Earth[49]). Finally, engineering improvements are required to support the risk taking and longevity that future science operations will require, such as *an ability to remove dust*, perhaps by a second robot "tender," (e.g., cleaning lenses and mirrors—dust completely obscured Opportunity's Mini-TES after the 2007 global dust storm[50]), *better safing methods* (e.g., detecting and handling navigation failures, low power, instrument calibration errors), and *closed-loop instrument placement* (reducing the need for iterative, multisol approaches, e.g., improving visual target tracking[51]).

If we could effectively make all of these improvements, and if urgency for studying Mars were not important, there might be no scientific reason for the scientists to be present on Mars to conduct their investigation. It would take longer to do the work remotely, and perhaps insights are hampered by the delay between deciding to do something on Mars and interpreting results, but the MER project provides compelling evidence that programmed surface robotic and orbital laboratories might be sufficient. Nevertheless, some technological and political questions remain.

Problems of autonomous mobility (as may be required for investigating caves), dexterity (e.g., hand-eye coordination that Phoenix sorely lacked), and scale (e.g., drilling kilometers into the surface) are technologically challenging and expensive. The history of mission development raises serious doubts about when all of the robotic capabilities required by scientists could be practically realized. Although no technological miracles are implied by the list of requirements, probably more academic research would need to shift from demonstrating what is technically possible (robotic hand-eye coordination was shown in research projects in the late 1980s) to practical design for extreme environments, miniaturization imposed by power and packaging constraints, and surface deployment packaging. The difficulties of practical systems engineering, illustrated by the delays and cost overruns of MSL (life-cycle costs have increased more than 50 percent over the original budget to more than \$2.5 billion[52]) and compounded by inherently long project development times and at best biannual projects, suggest that even with a slowly progressing human spaceflight capability, people might be on the martian surface before all of the scientifically required robotic technologies are launched.

Politics is another reason for suspecting that the future of Mars scientific exploration will be the story of people on the surface working with robotic laboratories. Future history is not determined by the logic of a scientist, engineer, or an accountant, but by society's preferences, values, and circumstances. Ironically, some scientists strongly support human spaceflight because they want to get answers about Mars in their lifetime. For individuals, time is a major consideration; they feel personal urgency for the program to move more quickly. Furthermore, although remotely controlled laboratories might be less expensive than human spaceflight, MSL's development record suggests that the future investments required for capable robotic systems could perhaps tilt the balance—sending people with heavy instruments and vehicles may be more appealing than encumbering generations of scientists to tediously manipulating tools from afar. Or a compromise might be possible in which astronauts teleoperate (joystick) robotic systems from Mars orbit, thus reducing automation requirements and improving the pace of the surveys and ability to drill and sample materials.[53]

Our exploration of Mars's surface is at its very beginning. Extrapolating from our experience with Spirit and Opportunity, we can conclude that covering the planet sufficiently with rovers like MER alone to prove that we've found everything of interest will require thousands of scientists and engineers working over many decades with at least dozens of rovers, costing at least tens of billions of dollars. With respect to exploring the entire planet, the MER missions are perhaps akin to the Norse voyages of 1000 AD to 1350 AD, which established a toehold in Newfoundland.[54] The exploration of Mars, like the investigations of North America's waterways and mineral deposits, will require a complex and varied package of vehicles, sensors, and traverses. The MER experience suggests that a range of rover sizes—including something between the size of MER and Sojourner and others larger than MSL—complemented by aerial and/or orbital reconnaissance, will be required for the detailed, relatively long duration studies of small areas like Home Plate and the relatively fast, long traverses between craters like those on Meridiani. In this respect, the proper analogy with ocean-faring voyages of Earth will occur only much later, as we circumnavigate Mars in multiyear missions, investigating areas previously visited and resolving remaining mysteries about its climate, landforms, minerals, and possible life.

What is the place for people in this mix of reconnaissance and scientific exploration? Satellites orbiting Earth have already demonstrated the possibility of automatically searching for features, reducing the effort of targeting and analyzing data.[54] One can fantasize about rovers with perceptual and conceptual abilities to survey areas, deploy instruments in a scientifically meaningful way, and report back about promising findings. Even when such "intelligent machines" do exist, people will still want to make their own interpretations and judgments about the exploration—after all, why would we have built these devices if not for our own purposes? At least in the short term, we need to focus on better engineering tools for automating the detailed programming of rovers and preparing received data for analysis rather than trying to convert remote laboratories into robotic geologists. The "autonomy" we need most will be obvious, bridging the gaps as we extend our reach and analysis—as Sims said, enabling people to do more of what they want to do.

Trebi-Ollennu reminds us that the problem is to design a scientific exploration *system*, which is not necessarily accomplished by putting all of the capabilities of people and machines into one component, a robot:

> The fascinating thing is *the human interaction with the robot*, which is based on the skills that both have. It's just like in a soccer team. Everybody has different skills. And you want to utilize each skill appropriately to optimize science return. That's the challenge. I don't think it's taken that way in spacecraft operations. What I have learned is that putting such autonomy on the rover or the spacecraft is not always the best way to get science return. There are certain things humans do really well. And just getting that balance I think is what has made us more successful on MER. Because we know what the rovers can do, and we know what we can do on the ground that can aid, to get better science return. And I think we do that really well.

As always when relating people and computer technology, the challenge is being clear about the relative roles of the robotic laboratory and the people. Focusing on the field science activity clarifies what we are trying to do and what we have accomplished. Where do we want to explore? What do we need to sample? What do we need to analyze in the samples? We can design better tools by understanding the work to be done on Mars, including the likely distribution of sites of interest (where are the eggs hidden?).

If it turns out that remotely reprogrammed systems can be made adequate for scientists' needs—fitting social urgency and budgets—then the articulated motivations for human spaceflight will need to shift from talk about scientific exploration to something more deliberately poetic, commercially practical, or frankly political and nationalistic. Until then, we can all share the Mars scientists' frustrations and joys in doing what would be a marvel to any of Earth's early explorers and is wondrous to any of us who stop to appreciate the adventure—the scientific exploration of the landscape of a distant planet.

Epilogue

When Squyres says, "We are conducting the first overland expedition across the surface of another planet, *ever*, in human history,"[1] he is making a scientific statement about the nature of the investigation while expressing an explorer's excitement of being the first to do it. For the scientists, the story of MER naturally weaves these two ways of thinking: one rigorously scientific, the other personal and sometimes broadly romantic. When justifying the space program, serious scientific, technological, economic, educational, and sometimes political rationalizations are publicly expected.[2] Lyrical and emotional inspirations are left for the entertainment world of science fiction literature, television, and movies, such as the *Star Trek* franchise.

My interest in this book has been to reveal the multiple dimensions of the scientists' work and their motivations, both scientific and personal. In closing, with a more poetic perspective, I want to return to the aesthetic and emotional aspects of being a scientist on a planetary mission and in particular the narrative of the "voyage of discovery" and the "intrepid explorer" that lies behind the personification of the rovers. Why does the myth of the hero on a journey fit MER so well? How might this romantic interpretation of a rover influence the public's interest and support for space exploration?

In May 2011, the Associated Press reported Spirit's demise: "Spirit, the scrappy robot geologist that captivated the world with its antics on Mars before getting stuck in a sand trap, is about to meet its end after six productive years." The melodrama of the rover is told here in the genre of a lost person ("incommunicado") with unabashed flourish: "As far as sibling rivalry went, Opportunity was the overachiever while Spirit was every bit the drama queen underdog." Neither tongue-in-cheek nor with signaled irony, the tone is amazing for assuming the audience's collusion: "The mission's deputy investigator, Ray Arvidson of Washington University in St. Louis, said he will remember Spirit as a fighter. 'It wouldn't quit, just like the little engine that chugged up the hill,' said Arvidson, referring to the children's bedtime story." It's easy to imagine these piled metaphors told with a smile, or even as jokes. But the humorous allusions—"unable to provide roadside assistance, NASA declared an end"—are dotted with what are after all serious self-descriptions of the mission team's character and its accomplishments: "The plucky rover will be remembered for demystifying Mars to the masses"; "This is a story of perseverance."[3]

As much as I have rallied for plain speaking and astute philosophical analysis, the poetic personification of MER is obviously no less real and important in our shared experience of this mission. "The rover as hero" has become an idiom, apparently infectious and for some tastes maybe overdone. But the poetry reveals how space exploration affects us emotionally and arguably reveals the essential aspect of human nature that will ultimately drive and sustain the effort and cost of voyages to the stars.

MER was preceded by many other "voyages of discovery" dramatized by historical and anthropomorphic allusions. In the space exploration literature, interplanetary probes have been portrayed as ships sailing into unsurveyed regions of the solar system. The "grand tour" missions of the Pioneers and Voyagers, launched in the 1970s, are well-known examples. Similarly, Meltzer claims that Galileo is "one of the greatest and most successful voyages of discovery ever attempted."[4] He describes how Galileo's anthropomorphic capabilities enabled scientists to explore the Jupiter system in the 1990s: "The Galileo spacecraft explored new, exciting territory, using instruments that were, in essence, the eyes, ears, and fingertips of humankind. As the Galileo Probe buried itself in Jupiter's atmosphere, as the Orbiter skimmed over Io's fire fountains and Europa's ice and possible buried ocean, it was we who were exploring uncharted frontiers."[5]

This description of Galileo combines the heroic "robot as explorer" metaphor with a resounding recognition that people were investigating Io and Europa through the virtual presence the spacecraft afforded ("it was we"). The pervasiveness and apparent value of the "voyage" metaphor to scientists themselves in framing their investigation is remarkable. For example, John Baross, a professor of oceanography, says, "Like Darwin's Beagle voyage, the current astrobiology 'voyage' in search for life throughout the universe promises to profoundly alter and expand our notions of life."[6]

As we learned from the MER scientists' emotional reactions to the rovers' landing, certain moments are experienced with a profound conception of the meaning of what we are accomplishing in these forays beyond our cradle. For example, astronaut William Anders, on returning from the first circumlunar flight of Apollo 8, reflected on being the first to see the earthrise: "It is ironic that we had come to study the moon, and it [the mission] was really discovering the Earth." And so he echoed T. S. Eliot's words, "We shall not cease from exploration, And the end of all our exploring, Will be to arrive where we started, And know the place for the first time."[7] This refrain would be a cliché, but for the cosmic discoveries of space science and astrobiology that continually give new perspective—in ungraspable magnitudes of distance, time, and improbability—on what it means to be on Earth, alive.

Personal stories from the astronauts and planetary scientists can help orient the space program and be part of its positive effect. Accordingly, some argue that human spaceflight is more interesting to the public than robotic missions, serving to sustain funding support and to inspire children to seek scientific and technical careers. For example, John Casani, the first Galileo project manager, made these comments on the importance of human missions: "Young people and older people alike … like heroes. What if instead of Admiral Byrd going to the South Pole, it was just an automated sled? It might have been able to do more science than he could, but it wouldn't have inspired the kind of interest and public reaction that [Byrd] did. Having heroes is an important part of any enterprise like this."[8]

All of our interests are personal interests, framed by social relations, and so we personalize spacecraft and identify with explorers. Correspondingly, how explorers are portrayed—as in the photographs of the mountain climber with a flag on a peak—symbolizes both personal accomplishment and social status. Dress—especially a formal uniform—marks a role and connotes power; it may be deliberately designed to evoke respect. For example, the introduction of the National Park ranger with a classic, flat-brim hat, personalized America's budding parks, focusing attention on an image of respected and idealized authority.[9] The 1960s photographs of astronauts in pressurized suits holding a helmet have a related social implication of personal stature with national import. Such idealized personas come to represent our cultural identity: we are a people who protect our remaining exploration places as National Parks. We are people who go into space.

The relation of adversity in harsh environments, heroism, and role models was eloquently expressed by Robert Zubrin, president of the Mars Society, at the inauguration of the research station on Devon Island in the Canadian Arctic in July 2000. He described how recovering from the setbacks of a failed airdrop, causing the loss of the habitat's floors and crane, "required skill, imagination, courage, team work, hard work, and above all grit and determination," and explained their social importance: "These are the qualities that will get us to Mars. Because of your heroism, our movement is going to grow. People will join it. They will join it to be on the same team with people like you."[10] Standing with Zubrin, with the orange evening sun brightening the habitat behind our small gathering, I could feel that tingling sense of historical place and being, of what we were doing, and share a tantalizing grasp of a barely imaginable future we were making, as Zubrin continued:

> Look how much things have changed since 1900. We now have cities with skyscrapers 100 stories tall and 200-ton ships flying through the air at 500 miles an hour across the oceans—we live in a science fiction world by the standards of the year 1900. Looking at this progress, who can imagine, that if we dare, that if we crack the shell, that if we break out of the cradle, who can deny that 1000 years from now that there cannot be a new branch of human civilization on Mars? And looking back over the past 1000 years, at the world lit only by fire, the cold world of early medieval times and how much things have changed since then, and how much magical almost our world would have to seem to someone from that time, who can deny that if we take this step, that if we break out of this cradle, that 1000 years from now that there won't be 1000's of branches of human civilization on 1000's of planets orbiting stars in this neck of the galaxy?[11]

And so portraying spacecraft or rovers as heroic personas—as projections of people—is an artistic way of representing our over-arching aspiration to extend our being and experience beyond the Earth, this belief that all we know and value might radiate and grow in space, that in the words of Spock, humanity might "live long and prosper."

A graphic of the Galileo spacecraft, titled "The Demise of Galileo," renders the spacecraft's streaming breakup on entering the Jovian atmosphere as a faint outline of a female spirit.[12] It's an unusual depiction, because boxy spacecraft in the void are not usually anthropomorphized. Our meditations on MER have been different not only because of the rover's body, but also because of its story. It sleeps, wakes up in the morning, strains to climb hills, slips and gets stuck, and provides far-ranging views. The rovers act like mechanical creatures. Even on the surface, the Phoenix lander was still called a "spacecraft," perhaps because it didn't move. Spirit and Opportunity are described as creatures not only because of their physical form, but because of the story of their overland travails.

Consider how well roving on Mars fits the general pattern of "the hero's journey," a narrative structure formalized by Joseph Campbell. This pattern, which resembles myths from many cultures throughout history, is used sometimes quite deliberately in literature, theater, and movies. For example, the famous *Star Wars* film series tells the quests of father and son from humble home planets to becoming heroes in galactic battles, all to discover the harmony and power that lies within "the force." Following the framework of such mythic plots, a space expedition naturally follows a familiar development of a hero on an adventure journey, with stages such as the Call to Adventure, the Crossing to a Special World, Ordeals, Rewards, and the Return with Elixir, which—befitting science fiction—is on MER reduced to physical parameters conveyed home in a stream of bits.[13] Zubrin's speech is the first part of the plot, the call for adventure, the call for humanity to break out of the cradle.

Our society's space exploration vision contains two often-competing fantasies of heroic adventure: people living and working in space (particularly on Mars) and robotic geologists. Ironically, with the robotic successes, some people will be even more motivated to go to Mars so that they can participate first-hand in making discoveries. Former NASA Administrator Griffin said, "I firmly believe that men and women putting footprints in the lunar dust and Martian sands, seeing and experiencing new worlds with our own eyes, is the essence of what makes us human."[14] But this diplomatic cliché of "human essence" doesn't adequately reveal the particular, psychological force we feel inside, the compulsion that moves us: you can't pin us down here—given the opportunity and the means, at least some people are leaving this planet for other places.

Aside from all the technical talk of trade-offs in doing scientific investigations or engineering assembly and repair work with robotic tools instead of astronauts, despite any arguments about how independent such technologies may become, the restless urge to "be there," to live and know a new place, will likely dominate any scientific, technical, economic, or political argument when the means become available and affordable. If it can be done at all, some people will go into space, as surely as some of our ancestors left Africa and covered every corner of this globe. When asked why we don't simply send robots to Mars when they eventually are more capable, rather than send people at greater expense sooner, Chris McKay, a NASA planetary scientist replied, "Even if computers progress so much that they can go to Paris, taste the wine, eat the food and come back and tell me all about it, I'd still want to go myself."[15]

It is this psychological reality that lies behind the talk of the "robotic geologist"—all that poetry and overdrawn metaphor expresses how the thought of being on Mars personally, for real, resonates emotionally and beckons for action. Projecting ourselves into the rovers is how we artistically express our urge *to be there*.

To be sure, the artful wrapping of the rovers' hard-earned accomplishments makes a nice way of telling the MER story—for the media reporters who strive to simplify, for the JPL mission chroniclers and scientists themselves whose nothing-but-the-facts technical reporting cloaks the team in anonymity, and for kids who can imagine their rover toys in sandboxes as being on another planet.

What the bureaucrats often feel obligated to ignore is that the human spirit cannot be partitioned into "serious" and "emotional"; rather, our psychic life is always poetic and rational at the same time. As the JPL MER reports attest, we can see the rovers' perambulations as a hero's journey, and this "seeing as" metaphor does not detract from the scientific work. The MER scientists have shown us as well in their projections of themselves into the rover and aesthetic choices that it is possible to "fuse the sensitivity of a poet with the rigor and self-discipline of a scientist."[16] It's okay to take a picture from the top of the Columbia Hills because "it's cool"—and it's okay to say that.

Very likely, scientists and engineers will someday go to Mars. People will be accompanied by rovers that both do their bidding and watch over them. On Mars, we will gladly operate rovers from within the warmth and safety of our base camps; we will ride the rovers up hills we would not want to climb anyway; we will lower rovers on cranes into steep ravines. We will carry them back for repair, and someday a rover might rescue one of us. For not every robot will be a laboratory or a transportation vehicle; some will dig ditches and others mindlessly inspect exposed equipment. Then, too, the poets and photographers remind us that not every member of the crew will be a scientist or engineer.

When people walk on Mars, we will look back on the lessons we learned from the MER—how people can collaboratively explore a new land, how in their diversity they find a niche and adapt their expertise, and how people learned to program a mobile laboratory to move, see, and detect the land for them. We will remember that they did this together—the scientists and engineers, the geologists and chemists, the old-timers and students—traveling as one team over several years in a journey shared with millions of awestruck people on TV and the Internet. We will remember this journey—our travel with resolute Spirit at Gusev and lucky Opportunity at Meridiani—as the first voyages of discovery on Mars.

Notes

Preface

1. Roxana Wales, Valerie Shalin, and Deborah Bass, "Requesting Distant Robotic Action: An Ontology for Naming and Action Identification for Planning on the Mars Exploration Rover Mission," *Journal of the Association For Information* 8, no. 2 (February 2007): 75–104.

2. Zara Mirmalek, "Solar Discrepancies: Mars Exploration and the Curious Problem of Interplanetary Time" (PhD diss., Department of Communication, University of California, San Diego, 2008).

3. William J. Clancey, "Clear Speaking about Machines: Scientists Are Exploring Mars, Not Robots" (paper presented at the Association for the Advancement of Artificial Intelligence Workshop: The Human Implications of Human-Robotic Interaction, Boston, 2006).

4. Sherry Turkle, with additional essays by William J. Clancey, Stefan Helmreich, Yanni A. Loukissas, and Natasha Myers, *Simulation and Its Discontents* (Cambridge, MA: MIT Press, 2009).

5. Steve Squyres, *Roving Mars: Spirit, Opportunity, and the Exploration of the Red Planet* (New York: Hyperion, 2005).

6. Steve W. Squyres et al., "The Spirit Rover's Athena Science Investigation at Gusev Crater, Mars," *Science* 305, no. 5685 (August 6, 2004): 794–799; Steve W. Squyres et al., "The Opportunity Rover's Athena Science Investigation at Meridiani Planum, Mars," *Science* 306 (December 3, 2004): 1698–1703; Steve W. Squyres et al., "Two Years at Meridiani Planum: Results from the Opportunity Rover," *Science* 313 (September 8, 2006): 1403–1407; Steve W. Squyres et al., "Exploration of Victoria Crater by the Mars Rover Opportunity," *Science* 324 (May 22, 2009): 1058–1061.

7. James K. Erickson, "Living the Dream," *IEEE Robotics & Automation Magazine* 13, no. 2 (June 2006), 12–18; for Sojourner, see Andy Mishkin, *Sojourner: An Insider's View of the Mars Pathfinder Mission* (New York: Berkley Publishing Group, 2003).

8. Edward C. Ezell and Linda N. Ezell, *On Mars: Exploration of the Red Planet: 1958–1978* (Washington, DC: National Aeronautics and Space Administration Special Publication-4212, 1984), xvi.

1 Scientists Working on Mars

1. A Mars day, called a sol, has 24 hours, and each Mars hour has 60 minutes, as on Earth. But Mars rotates more slowly, so the 24-hour clock on Mars takes longer to complete a cycle. By choice, rather than saying an hour has more minutes or a minute has more seconds, scientists have decided that a Mars second simply takes longer. If your clock designed for Mars and one designed for Earth are side by side, the Mars clock will appear to take longer to tick. The nature and implications of "Mars time" are examined in great detail by Mirmalek, "Solar Discrepancies."

2. At the time of this writing, in 2012, Opportunity is still operating on Mars and the practices described here continue. I generally refer to events during the first ninety sols, the nominal mission, using the past tense.

3. Jim Bell, "Photographing Mars," *Planetary Report* 26, no. 6 (November–December 2006): 12–18.

4. A "sequence" is an ordered list of commands for controlling the rover. The scientists' informal "observations" (e.g., "Pancam surface texture of Merlot Krispies [soil target]") are converted by engineers using software into instructions the MER rover can process. Thus, informally "sequence" refers to these commands at any level of abstraction.

5. MER field notes, NASA History Division file "04.02.18 M25 SCIENCE wjc.doc." File names use the format used in the MER project: "YY.MM.DD <M or G><sol number> SCIENCE <author initials>," where M means Meridiani and G means Gusev. "MER A" is also used to refer to Spirit; "MER B" to Opportunity.

6. MER field notes, "04.02.18 M25 SCIENCE wjc.doc."

7. The distinction between professional knowledge as tacit skill versus articulated theory comes from Donald A. Schön, *Educating the Reflective Practitioner: Toward a New Design for Teaching and Learning in Professions* (San Francisco: Jossey-Bass, 1987). In practice, most conceptions and actions are routine and not mediated by justifications and inferential arguments. Chapter 7, "The Communal Scientist," develops these ideas in detail.

8. See the bibliographical essay for references related to the nature of work practice.

9. See JPL MER Mission Updates for archived status and distance data: http://marsrovers.jpl. nasa.gov/home/index.html.

10. JPL Press Release 2004-003.

11. JPL MER Mission Updates, "Sol 778–783, Mar 16, 2006: Spirit Continues Driving on Five Wheels," http://marsrovers.jpl.nasa.gov/mission/status_spiritAll_2006.html#sol778 (accessed March 20, 2006).

12. Ibid.

13. NASA 2000; italics added.

14. Matt Golombek, "Spirit and Opportunity—Martian Geologists," *Planetary Report* 27, no. 3 (May–June 2007): 12–16.

15. Ibid., 13.

16. Ibid., 16.

17. Andrew Chaikin, *A Passion for Mars: Intrepid Explorers of the Red Planet* (New York: Abrams, 2008).

18. Ezell and Ezell, *On Mars*, 337.

19. Chaikin, *A Passion for Mars*, 258.

20. Visualizing the work area and planning the investigation focuses on one robot at a time; thus, for simplicity I refer to "the rover" throughout. Of course, there were two rovers involved with two teams of scientists and engineers managing them and thus in some sense two MER missions. Broader organizational issues related to coordinating group and individual attention and other resources between the two rovers are relevant but are not considered here. In general, individuals were dedicated to one team throughout the nominal mission; however, some scientists—particularly the principal investigator—shifted between rovers according to what was happening at each site. Some engineers more routinely shifted between rovers as needed after the nominal mission.

2 Mission Origin and Accomplishments

1. Chris McKay, "The Case for Mars 2009," *The Mars Quarterly* 1, no. 2 (Mars Society, 2009): 4.

2. Steve Squyres, quoted by Chris Carberry, "Interview with Dr. Steve Squyres," *The Mars Quarterly* 1, no. 2 (Mars Society, 2009): 12–14.

3. Golombek, "Spirit and Opportunity—Martian Geologists," 12–16.

4. Carberry, "Interview with Dr. Steve Squyres," 12.

5. Squyres et al., "Investigation at Meridiani Planum."

6. Squyres, *Roving Mars*.

7. NASA Headquarters Press Release, "NASA Outlines Mars Exploration Program for Next Two Decades (October 26, 2000)," http://sse.jpl.nasa.gov/news/display.cfm?News_ID=493 (accessed November 19, 2011).

8. Andrew Fazekas, "Making Tracks: Robotic Rovers Roll through Space and Time," *Astronomy* (November 17, 2004), http://www.astronomy.com/asy/default.aspx?c=a&id=2599 (accessed July 27, 2009).

9. Andrew Chaikin, "The Other Moon Landings," *Air & Space* (March 1, 2004): 31–37. See also Wikipedia, "Lunokhod 1," http://en.wikipedia.org/wiki/Lunokhod_1 (accessed January 31, 2008) and National Space Science Data Center, "Luna 21/Lunokhod 2," http://nssdc.gsfc.nasa.gov/nmc/masterCatalog.do?sc=1973-001A (accessed February 14, 2008).

10. Ezell and Ezell, *On Mars*, 419, quoting the Gerald Soffen Dryden lecture at the American Institute of Aeronautics and Astronautics' 16th Aerospace Science Meeting in Huntsville, Alabama, January 1978.

11. JPL MER Mission, "Baby Boomers on Mars," http://marsrovers.jpl.nasa.gov/spotlight/20060320.html (accessed December 1, 2007).

12. Unless noted otherwise, all quotes in the text from Cabrol, Carr, Des Marais, Matijevic, Rice, Sims, Squyres, Trebi-Ollennu, and Yingst are from the author's interviews. See chapter 5, note 3, for details.

13. Mishkin, *Sojourner*, 234–243.

14. Ibid., 305.

15. Squyres, *Roving Mars*, 81, 378.

16. Steve Squyres, quoted by Chaikin, *A Passion for Mars*, 244.

17. Squyres, *Roving Mars*, 81.

18. JPL MER Mission, "The Rover's 'Eyes' and Other 'Senses,'" http://marsrover.nasa.gov/mission/spacecraft_rover_eyes.html (accessed November 24, 2011).

19. Philip R. Christensen et al., "The Miniature Thermal Emission Spectrometer for the Mars Exploration Rovers," *Journal of Geophysical Research* 108, no. E12 (2003): 8064.

20. Darby Dyar, "An Out-of-this-World Interview with Darby Dyar," *College Street Journal*, December 12, 2002, http://www.mtholyoke.edu/offices/comm/csj/121302/dyar.shtml (accessed June 25, 2009).

21. JPL MER Mission, "NASA's Mars Rover Finds Evidence of Ancient Volcanic Explosion," May 3, 2007, http://marsrovers.jpl.nasa.gov/newsroom/pressreleases/20070503a.html (accessed May 20, 2007).

22. JPL MER Mission, "Press Release Images: Opportunity," http://marsrovers.nasa.gov/gallery/press/opportunity/20040503a.html (accessed January 24, 2008).

23. JPL MER Mission, "Now a Stationary Research Platform, NASA's Mars Rover Spirit Starts a New Chapter in Red Planet Scientific Studies," http://marsrovers.jpl.nasa.gov/newsroom/pressreleases/20100126a.html (accessed January 26, 2010).

24. JPL MER Mission, "Spirit May Have Begun Months-Long Hibernation," http://marsrovers.jpl.nasa.gov/newsroom/pressreleases/20100331a.html (accessed March 31, 2010).

25. Michael Sims and David Des Marais, email messages to author, July 18, 2011.

3   A New Kind of Field Science

1. Robert E. Kohler, *Landscapes and Labscapes: Exploring the Lab-Field Border in Biology* (Chicago: University of Chicago Press, 2002), 10, referencing Bruce Hevly, "The Heroic Science of Glacier Motion," *Osiris* 11 (1996): 66–86.

2. Martin Rudwick, "Geological Travel and Theoretical Innovation: The Role of 'Liminal' Experience," *Social Studies of Science* 26 (1996): 143–159.

3. Kohler, *Landscapes and Labscapes*, 6.

4. Martin Rudwick, *The New Science of Geology* (Surrey, UK: Ashgate Variorum, 2004); Rudwick, "Geological Travel and Theoretical Innovation," 144–145.

5. For example, see Pascal Lee, Charlie Cockell, and Chris McKay, "Gullies on Mars: Origin by Snow and Ice Melting and Potential for Life Based on Possible Analogs from Devon Island, High Arctic," *35th Lunar and Planetary Science Conference*, abstract no. 2122 (2004).

6. Jonathan Clarke, ed., *Mars Analog Research*, vol. 111, American Astronautical Society Science and Technology Series, AAS 06–263 (San Diego: Univelt, Inc., 2006).

7. "Haughton–Mars Project," http://www.marsonearth.org (accessed July 4, 2003).

8. NASA Ames Press Release, "NASA Field-Tests the First System Designed to Drill for Subsurface Martian Life," July 5, 2005, http://www.nasa.gov/centers/ames/research/exploringtheuniverse/marsdrill.html (accessed July 27, 2009). See also "Mars Astrobiology Research and Technology Experiment," http://marte.arc.nasa.gov (accessed July 27, 2009).

9. Darlene Lim et al., "Limnology of Pavilion Lake B.C.: Characterization of a Microbialite Forming Environment," *Fundamental and Applied Limnology* 173, no. 4 (2009): 329–351. See also "Pavilion Lake Research Project," http://www.pavilionlake.com (accessed July 27, 2009).

10. NASA Ames Press Release, "NASA Astrobiologists to study extreme life at Earth's highest lake," October 10, 2002, http://www.nasa.gov/centers/ames/news/releases/2002/02_109AR.html (accessed July 27, 2009). See also http://www.extremeenvironment.com/2002/team/cabrol.htm (accessed November 20, 2011).

11. William "Red" Whittaker, Deepak Bapna, Mark W. Maimone, and Eric Rollins, "Atacama Desert Trek: A Planetary Analog Field Experiment (1997)," http://www.cs.cmu.edu/afs/cs/project/lri-13/www/atacama-trek/conference_paper/i-sairas3.html (accessed June 27, 2009).

12. Henry Bortman, "Probing Antarctica's Lake Bonney," *Astrobiology Magazine*, 21 May 2009, http://www.astrobio.net/index.php?option=com_expedition&task=detail&id=3130 (accessed November 20, 2011).

13. William J. Clancey, "Participant Observation of a Mars Surface Habitat Simulation," *Habitation: International Journal for Human Support Research* 11, no. 1/2 (2006): 27–47.

14. Unless noted otherwise, all quotations in the text from Cabrol, Carr, Des Marais, Matijevic, Rice, Sims, Squyres, Trebi-Ollennu, and Yingst are from the author's interviews. See chapter 5, note 3, for details.

15. FIDO lacked the RAT (performed manually by field support personnel), APXS (nonfunctional in 1 bar), and Mössbauer Spectrometer (a radiation hazard); the team was provided with related spectrometry data from this region. See Edward Tunstel et al., "FIDO Rover Field Trials as Rehearsal for the NASA 2003 Mars Exploration Rovers Mission," *Automation Congress, 2002 Proceedings of the 5th Biannual World* 14 (2002): 320–327. See also JPL Memo, "Mars Exploration Rover Mission: Rover Test Plan for Mars Yard and Field Trials—Spring 2001," draft report 0.4 (March 27, 2001).

16. Ezell and Ezell, *On Mars*, 258.

17. For an engineering description of FIDO test configurations, see Edward Tunstel, "Prototype Rover Field Testing and Planetary Surface Operations," *Proceedings of the Performance Metrics for Intelligent Systems Workshop* (PerMIS'07) (Washington, DC: ACM, 2007), 196–203.

18. Kohler, *Landscapes and Labscapes*, 2.

19. Carol Stoker, "Science Strategy for Human Exploration of Mars," in *Strategies for Mars: A Guide to Human Exploration*, ed. Carol Stoker and Carter Emmart, vol. 86, Science & Technology Series (San Diego: Univelt, 1996), 537–560; quotation cited 540–541.

20. Kohler, *Landscapes and Labscapes*, 3.

21. Ibid., 98.

22. For early history of biological fieldwork, see ibid., 114–115.

23. JPL MER Mission, "What's in a Name? It Depends on Who's Doing the Naming," http://marsrovers.jpl.nasa.gov/spotlight/spirit/a24_20040602.html (accessed June 16, 2004). See also A. J. S. Rayl, "Mars Exploration Rovers Update: Spirit Gets Back Home (To Where It Once Belonged), Opportunity Completes 10K at Victoria's Rim," *The Planetary Society*, February 28, 2007, http://www.planetary.org/news/2007/0228_Mars_Exploration_Rovers_Update_Spirit.html (accessed January 24, 2008).

24. Space.com, "Mars Rover Sees Huge Crater Better than Ever," http://www.space.com/8686-mars-rover-sees-huge-crater.html (accessed July 18, 2011).

25. "List of Craters on Mars," http://en.wikipedia.org/wiki/List_of_craters_on_Mars (accessed July 18, 2011).

26. NASA, "Liftoff to Learning: Voyage of Endeavour—Then and Now," http://www1.nasa.gov/audience/foreducators/topnav/schedule/programdescriptions/Voyage_of_Endeavour_5-12.html (accessed July 28, 2009). Video available online: http://quest.nasa.gov/content/rafiles/space/voyage.rm.

27. For sociopolitical interpretations of these voyages, see Patricia Fara, *Science: A Four Thousand Year History* (New York: Cambridge University Press, 2009), 151; also Paul Mapp, "Silver, Science, and Routes to the West: The Pacific Ocean and Eighteenth-Century French Imperial Policy," *Common-Place* 5, no. 2 (January 2005), http://www.historycooperative.org/journals/cp/vol-05/no-02/mapp/index.shtml (accessed February 29, 2008).

28. BBC, "BBC History: Endeavour's Scientific Impact (1768–1771)," http://www.bbc.co.uk/history/british/empire_seapower/endeavour_voyage_01.shtml (accessed July 24, 2008).

29. JPL, "NASA Spacecraft Sees Ice on Mars Exposed by Meteor Impacts," http://www.spaceref.com/news/viewpr.html?pid=29232 (accessed September 25, 2009).

30. Malcolm Nicolson, "Historical Introduction," in Alexander von Humboldt, *Personal Narrative of Journey to the Equinoctial Regions of the New Continent* (London: Penguin Books, [1834] 1995), ix–xxiv; quotation cited x.

31. Ibid., xi–xii.

32. Ibid., lxii.

33. Ibid., xi.

34. Ibid.

35. Ibid., xiii.

36. Ibid., xxiii.

37. Ibid., xxiv.

38. Ibid.

39. Quoted by Nicolson, "Historical Introduction," lxii.

40. Golombek, "Spirit and Opportunity—Martian Geologists," 13.

41. Louis Friedman, "Too Much Pork, Too Little Exploration," The Planetary Society: Lou's View, September 29, 2010, http://www.planetary.org/programs/projects/space_information/20100929.html (accessed November 1, 2010).

42. The Planetary Society, "Beyond the Moon: A New Roadmap for Human Space Exploration in the 21st Century," November 2008, 7, http://planetary.org/special/roadmap/beyond_the_moon.pdf (accessed October 26, 2010).

43. Anthony Wilden, *The Rules Are No Game: The Strategy of Communication* (London: Routledge & Kegan Paul, 1987).

44. Wiesner Committee, "Report to the President-Elect of the Ad Hoc Committee on Space," January 10, 1961, http://www.hq.nasa.gov/office/pao/History/report61.html (accessed July 31, 2011).

45. Ezell and Ezell, *On Mars*.

46. Christopher Scolese, statement before Congressional Committee on Appropriations, April 29, 2009.

47. Steven J. Dick, "Exploration, Discovery, and Science," NASA History Division: Why We Explore, no. 11 (2005), http://history.nasa.gov/Why_We_/Why_We_11.html (accessed April 10, 2009).

48. William R. Corliss, *The Viking Mission to Mars* (Washington, DC: National Aeronautics and Space Administration, Special Publication-334, 1974).

49. Freeman Dyson, *The Scientist as Rebel* (New York: New York Review of Books, 2006), 227.

50. Janet A. Vertesi, "Seeing Like a Rover: Images in Interaction on the Mars Exploration Mission" (PhD diss., Science and Technology Studies Department, Cornell University, 2009), 216.

51. Squyres et al., "Exploration of Victoria Crater by the Mars Rover Opportunity."

52. White House, "President Bush Announces New Vision for Space Exploration Program," press release, January 14, 2004, http://www.spaceref.com/news/viewpr.html?pid=13404 (accessed January 14, 2004). See also NASA, "The New Age of Exploration: NASA's Direction for 2005 and Beyond," no. NP-2005–01–397-HQ, http://www.nasa.gov/pdf/107490main_FY06_ Direction.pdf (accessed February 7, 2005).

53. Dick, "Exploration, Discovery, and Science."

54. Stephen J. Pyne, "Seeking Newer Worlds: An Historical Context," in *Critical Issues in the History of Spaceflight*, ed. Steven J. Dick and Roger D. Launius (Washington, DC: National Aeronautics and Space Administration Special Publication-4702, 2006), 7–36; quotation cited 30.

55. Michael Reidy, Gary Kroll, and Erik Conway, *Exploration and Science: Social Impact and Interaction* (Oxford: ABC-CLIO, 2006), 269.

56. Richard A. Kerr, "Mars Rover Trapped in Sand, But What Can End a Mission?" *Science* 324 (May 22, 2009): 998.

57. Chaikin, *A Passion for Mars.*

58. Chaikin, personal communication, July 7, 2010.

59. JPL, "FIDO Field Test," http://marsrovers.jpl.nasa.gov/fido/humanrover.html (accessed July 18, 2011).

60. NASA/JPL, "People Are Robots, Too. Almost," *ScienceDaily* (October 29, 2003), http://www.sciencedaily.com/releases/2003/10/031029063517.htm (accessed May 15, 2009).

61. Ray Kurzweil, *The Singularity Is Near: When Humans Transcend Biology* (New York: Penguin, 2005).

62. The Mars Society Annual Convention, Washington, DC, August 2006.

63. Clancey, "Clear Speaking about Machines."

64. George Orwell, "Politics and the English Language," *Horizon* 13, no. 76 (April 1946): 252–265.

65. Edward Tufte, *Beautiful Evidence* (Cheshire, CT: Graphics Press, 2006), 75, 181, 162–168.

66. G. Scott Hubbard, Louis Friedman, and Kathryn Thornton, "Examining the Vision for Space Exploration: Workshop Findings and Roadmap Analysis," *59th International Astronautical Congress* (Glasgow, September 29, 2008), paper IAC-08-B3.1.6.

67. Steven J. Dick and Keith Cowing, *Risk and Exploration: Earth, Sea, and the Stars* (Washington, DC: National Aeronautics and Space Administration Special Publication-4701, 2005), http://history.arc.nasa.gov/risk+exploration.htm (accessed July 31, 2005).

68. Reidy, Kroll, and Conway, *Exploration and Science*, xi.

69. Pyne, "Seeking Newer Worlds," 13.

70. Robert Zubrin, "How to Go to Mars Right Now!" *IEEE Spectrum* 46, no. 6 (June 2009): 48–49, http://spectrum.ieee.org/aerospace/space-flight/how-to-go-to-marsright-now/3 (accessed June 3, 2009).

71. William J. Clancey, *Situated Cognition: On Human Knowledge and Computer Representations* (New York: Cambridge University Press, 1997). For a review of the constructionist

perspective, see William J. Clancey, "Scientific Antecedents of Situated Cognition," in *Cambridge Handbook of Situated Cognition*, ed. Philip Robbins and Murat Aydede (New York: Cambridge University Press, 2008), 11–34. See also Brendan Wallace, Ross, A., Davies, J. B., and Anderson T., eds., *The Mind, the Body and the World: Psychology after Cognitivism* (London: Imprint Academic, 2007).

72. For example, see Stephen J. Pyne, *Voyager: Seeking Newer Worlds in the Third Great Age of Discovery* (New York: Viking, 2010), 37–47.

73. Squyres presented at the MER press conference, "First Results from Spirit's Bedrock Analysis at Columbia Hills," August 18, 2004, Press Conference Video Archive, http://www.nasa.gov/vision/universe/solarsystem/MER_Video_Archive.html (accessed June 22, 2009). See also JPL MER Mission, "Spirit Probes Deeper into 'Clovis' site," August 23, 2004, http://marsrovers.jpl.nasa.gov/mission/status_spiritAll_2004.html#sol205 (accessed June 22, 2009).

74. McKay, "The Case for Mars 2009." A localized deposit is described by Oliver Morton, "Carbonate Deposits Found on Mars," *Nature News*, December 19, 2008, http://www.nature.com/news/2008/191208/full/news.2008.1329.html (accessed December 24, 2009). See also JPL MER, "NASA Rovers Finds Clue to Mars' Past and Environment for Life," June 3, 2010, http://marsrovers.jpl.nasa.gov/newsroom/pressreleases/20100603a.html (accessed June 15, 2010) and A. J. S. Rayl, "Mars Exploration Rovers Special Update: Spirit Team Announces Major Water Discovery," *Planetary News*, June 17, 2010, http://planetary.org/news/2010/0617_Mars_Exploration_Rovers_Special_Update.html (accessed July 26, 2011).

75. David L. Chandler, "Mars Rover's Disability Leads to Major Water Discovery," *New Scientist* 14, no. 49 (May 23, 2007), http://www.newscientist.com/article/dn11914-mars-rovers-disability-leads-to-major-water-discovery.html (accessed June 22, 2009).

76. William J. Clancey, *Conceptual Coordination: How the Mind Orders Experience in Time* (Mahwah, NJ: Lawrence Erlbaum, 1979).

77. Although human conceptual systems have been *modeled* in computer programs (notably scientific models), no computer program is capable of conceptualization, the higher-order temporal categorization of perceptual-motor organizations involved in procedural planning and modeling, which blends values and desires with possibly conflicting goals and actions. In turn, building tools for people or seeking to replace them with robots involves understanding the nature of conceptualization, including consciousness, more generally, and how identity drives and guides personal projects, more specifically. I have written two books explaining these ideas and several papers, as cited in the bibliographical essay. For an introduction based on examples from Apollo and analog field science, see William J. Clancey, "Roles for Agent Assistants in Field Science: Personal Projects and Collaboration," *IEEE Transactions on Systems, Man, and Cybernetics, Part C: Applications and Reviews* 34, no. 2 (2004): 125–137.

4   A New Kind of Scientific Exploration System

1. Steven Squyres, "Science Results from the Mars Exploration Rover Mission," NASA Ames Director's Colloquium, May 15, 2009.

2. The conflicts and revisions are listed in Squyres's formal response to PDR issues at the Critical Design Review; see JPL, "MER Mission System Critical Design Review," unpublished, December 11, 2001.

3. Roger D. Launius and Howard E. McCurdy, *Robots in Space* (Baltimore: Johns Hopkins University Press, 2008), xviii.

4. Ken Goldberg, ed., *The Robot in the Garden: Telerobotics and Telepistemology in the Age of the Internet* (Cambridge, MA: MIT Press, 2000), 7; Thomas B. Sheridan, "Teleoperation, Telerobotics, and Telepresence: A Progress Report," *Control Engineering Practice* 3, no. 2 (1995): 205–214; quotation cited 205.

5. Zarya: Soviet, Russian and International Spaceflight, "Luna 17: Carrier for Lunokhod 1," http://www.zarya.info/Diaries/Luna/Luna17.php (accessed August 24, 2009).

6. Stoker, "Science Strategy for Human Exploration of Mars," 544.

7. James S. Martin and A. Thomas Young, "Viking to Mars: Profile of a Space Exploration," *Astronautics & Aeronautics* 14 (November 1976): 22–42, 44–48; B. Gentry Lee, "Mission Operations Strategy for Viking," *Science* 194, no. 4260 (October 1976): 59–62.

8. A. Thomas Young, "Viking Mission Operations," in *Technology Today for Tomorrow: Proceedings of the Twelfth Space Congress* (Cocoa Beach, FL: Canaveral Council of Technical Societies, April 9–11, 1975), A75-40601 20-12, 6-41–6-48.

9. Interview with Jim Martin, March 14, 2001, by telephone with Jay Trimble and Roxana Wales; confidential notes provided by Jay Trimble.

10. Ezell and Ezell, *On Mars,* chapter 11.

11. Martin and Young, "Viking to Mars," 41.

12. Young, "Viking Mission Operations," 6–43.

13. Ezell and Ezell, *On Mars*, 396, 398. Regarding the boom, see 393–394. Regarding overnight operations, see Lee, "Mission Operations Strategy for Viking."

14. Mark Washburn, *Mars at Last!* (New York: Putnam, 1977), 226.

15. Joy A. Crisp, Mark Adler, Jacob R. Matijevic, Steven W. Squyres, Raymond E. Arvidson, and David M. Kass, "Mars Exploration Rover Mission," *Journal of Geophysical Research* 108, no. E12, 8061 (2003): 2-1–2-17.

16. Sheridan, "Teleoperation, Telerobotics, and Telepresence," 210.

17. Stoker, "Science Strategy for Human Exploration of Mars," 545.

18. For a review of underwater archeological field science using telerobotics, see David Mindell and Brian Bingham, "New Archaeological Uses of Autonomous Undersea Vehicles," *IEEE/MTS Oceans Conference* (November 2001).

19. Walter A. Aviles, "Telerobotic Remote Presence: Achievements and Challenges," in *Human Machine Interfaces for Teleoperators and Virtual Environments*, ed. Nathaniel I. Durlach, Thomas B. Sheridan, and Stephen R. Ellis (NASA CP 100071, 1990), 31; see also Sheridan, "Teleoperation, Telerobotics, and Telepresence," 207.

20. For example, see Thomas B. Sheridan, "Musings on Telepresence and Virtual Presence," *Presence* 1, no. 1 (1992): 120–126; Sheridan, "Teleoperation, Telerobotics, and Telepresence," 210; and Goldberg, *The Robot in the Garden*.

21. Michael W. McGreevy, "An Ethnographic Object-Oriented Analysis of Explorer Presence in a Volcanic Terrain Environment," NASA TM-108823 (Ames Research Center, Moffett Field, CA, 1994).

22. Eric M. De Jong et al., "The Benefits of Virtual Presence in Space (VPS) to Deep Space Missions," *AIAA 9th International Conference on Spacecraft Operations (SpaceOps)* (Rome, June 19–24, 2006), http://hdl.handle.net/2014/39866 (accessed July 19, 2009).

23. Pyne, "Seeking Newer Worlds," 34.

24. Vertesi, "Seeing Like a Rover: Images in Interaction on the Mars Exploration Mission," chapter 4.

25. Stoker, "Science Strategy for Human Exploration of Mars."

26. Gentry Lee, "Observations from Mars Exploration Rover Mission System PDR," memo to Robert Mitchell at JPL, unpublished, April 1, 2001; emphasis added.

27. Ibid.

28. Ezell and Ezell, *On Mars*, 360.

29. Mishkin, *Sojourner*, 235.

30. Interview with Matt Golombek, March 22, 2001, at JPL with Jay Trimble and Roxana Wales; confidential field notes provided by Jay Trimble.

31. Lee, "Mission Operations Strategy for Viking," 60.

32. Corliss, *The Viking Mission to Mars*, 74.

33. Mishkin, *Sojourner*, 297.

34. Interview with Carol Stoker, February 14, 2001, at NASA Ames, with Rich Keller, Jay Trimble, and Roxana Wales; quotes from field notes by Jay Trimble.

35. Mishkin, *Sojourner*, 95.

36. Interview with Matt Golombek, March 22, 2001.

37. Interview with Carol Stoker, February 14, 2001.

38. Arthur Amador, "Introduction to Operations Scenarios and Processes," unpublished (Pasadena, CA: JPL, November 2, 2000).

39. Justin V. Wick, John L. Callas, Jeffrey S. Norris, Mark W. Powell, and Marsette A. Vona, "Distributed Operations for the Mars Exploration Rover Mission with the Science Activity Planner," *Proceedings IEEE Aerospace Conference* (March 5–12, 2005), 4162–4173.

40. Steven W. Squyres et al., "Athena Mars Rover Science Investigation," *Journal of Geophysical Research* 108, no. E12, 8062 (2003): 3-1–3-21; quotation cited 3-3.

41. Ibid., 3-16.

42. JPL, "MER Surface Operations: Approach and Issues," *MER Surface Operations Workshop*, unpublished (Pasadena, CA: JPL, January 10, 2001), 16.

43. Christopher G. Salvo, "Uplink Process Design: MER MOS/GDS System Design Peer Review," unpublished (Pasadena, CA: JPL, October 26, 2001), 7.

44. JPL, "MER Surface Operations: Approach and Issues," *MER Surface Operations Workshop* (Pasadena, January 10, 2001), p. 19.

45. JPL, "MER Surface Operations," 17; see also Andrew Mishkin, "The MER Mission Operations System," unpublished (Pasadena, CA: JPL, September 5, 2002).

46. Mishkin, *Sojourner*, 292–293.

47. JPL, "MER Surface Operations," 19.

48. Ibid.

49. Squyres et al., "Athena Mars Rover Science Investigation," 15.

50. JPL, "MER MS PDR Board Report," CAS-RTM-0411.1, unpublished, April 11, 2011.

51. Ibid.

52. JPL, "MER Mission System Critical Design Review," 7.

53. Mishkin, "The MER Mission Operations System."

54. Bell, "Photographing Mars," 13.

5   The Mission Scientists

1. NASA Mars Exploration Program, "NASA Selects 28 Participating Scientists for Mars Rover Mission," May 30, 2002, http://marsprogram.jpl.nasa.gov/newsroom/pressreleases/20020530a. html (accessed July 23, 2009).

2. NASA, Mars Exploration Rover Mission, "NASA Expands Rover Science Team: October 19, 2005," http://marsrover.nasa.gov/spotlight/20051019.html (accessed July 24, 2009); this article states that forty-nine were selected previously, not fifty.

3. The MER scientists interviewed are Nathalie A. Cabrol, Michael H. Carr, David J. Des Marais, James W. Rice, Michael H. Sims, Steve Squyres, and R. Aileen Yingst. The MER engineers interviewed are Jake Matijevic and Ashitey Trebi-Ollennu. The interviews were conducted August 8–27, 2006 (at NASA Ames: Sims, Des Marais, and Cabrol; by telephone: Rice and Yingst; in Woodside, CA: Carr) and December 5–18, 2007 (at JPL: Squyres and Matijevic; by telephone: Trebi-Ollennu). The original recorded interviews and complete transcripts are filed with the NASA History Division as part of its Historical Reference Collection. Quotations in this monograph have been edited from the original conversations for clarity and readability, following suggestions from the speakers, with their review and approval. Clarifying phrases appear in square brackets [like this]. Comments about laughter and inflection appear in parenthesized italic remarks *(like this)*. Words or phrases emphasized in speech appear in italic. Omitted phrases and sentences are generally not indicated in the quotations; ellipses (…) usually mark contemplative pauses. Unless noted otherwise, all quotations in the text from Cabrol, Carr, Des Marais, Matijevic, Rice, Sims, Squyres, Trebi-Ollennu, and Yingst are from the interviews.

4. Squyres, *Roving Mars*, parts 2 and 3 describe many people and how they contributed to the design of instruments and JPL's operations.

5. Erickson, "Living the Dream," 12.

6. For additional interviews and analyses, see Vertesi, "Seeing Like a Rover: Images in Interaction on the Mars Exploration Mission," and Mirmalek, "Solar Discrepancies." An excellent set of research questions for studying such groups is provided by David Hackett Fischer, *Historians' Fallacies: Toward a Logic of Historical Thought* (New York: Harper Perennial, 1970), 241.

7. Joy Crisp Home Page, http://joycrisp.com (accessed June 17, 2009); Mark Whalen, "High Spirit, Great Opportunity," *Universe* (JPL, Office of Communications and Education, July 18, 2003), 3; Joy Crisp, email messages to author, June 17 and 19, 2009; Cornell, "Athena Mars Exploration Rovers: Way Cool Scientist," http://athena.cornell.edu/kids/cs_crisp.html (accessed June 16, 2009); NASA Mars Exploration Program, "Dr. Joy Crisp, Project Scientist for the Mars Exploration Rover Project," http://mars.jpl.nasa.gov/spotlight/joyCrisp.html (accessed October 28, 2004).

8. Squyres, *Roving Mars*, 31.

9. Kohler, *Landscapes and Labscapes*, 175.

10. Michael H. Carr, *The Surface of Mars* (New York: Cambridge University Press, 2007), ix.

11. Squyres, *Roving Mars*, 1–5.

12. Carr, *The Surface of Mars,* 231–254.

6   Being the Rover

1. Unless noted otherwise, all quotations from Cabrol, Carr, Des Marais, Matijevic, Rice, Sims, Trebi-Ollennu, and Yingst are from the interviews cited in chapter 5, note 3.

2. Squyres, *Roving Mars,* 328, 334–336.

3. Chaikin, *A Passion for Mars*, 251.

4. Here and throughout, if a quotation from Squyres is not marked by a note, it comes from the interview on December 5, 2007.

5. As Squyres reports in *Roving Mars*, 250, Mike Malin's group first compared older Mars Global Surveyor images with images from the Descent Image Motion Estimation System (DIMES) to locate the general landing area. Multimission Imaging Processing Lab (MIPL) then identified horizon landmarks in the initial Navcam images to more precisely locate the rover. See Malin Space Science Systems, "MER Activities at Malin Space Science Systems," http://www.msss.com/all_projects/mars-exploration-rovers.php (accessed November 23, 2011). See also Tim Parker et al., "Localization, Localization, Localization," *35th Lunar and Planetary Science Conference*, abstract no. 2189 (2004).

6. Vertesi, "Seeing Like a Rover: Images in Interaction on the Mars Exploration Mission," 176.

7. Ezell and Ezell, *On Mars*, 413.

8. Squyres et al., "Exploration of Victoria Crater by the Mars Rover Opportunity."

9. Chaikin, *A Passion for Mars*, 248.

10. Jeffrey S. Norris, Mark W. Powell, Jason M. Fox, Kenneth J. Rabe, and I-Hsiang Shu, "Science Operations Interfaces for Mars Surface Exploration," *2005 IEEE International Conference Systems, Man and Cybernetics* 2 (October 12, 2005), 1365–1371; Marsette A. Vona, Paul G. Backes, Jeffrey S. Norris, and Mark W. Powell, "Challenges in 3D Visualization for Mars Exploration Rover Mission Science Planning," *Proceedings IEEE Aerospace Conference* 8 (March 8–15, 2003), 8-3541–8-3551.

11. Erickson, "Living the Dream"; Wick et al., "Distributed Operations for the Mars Exploration Rover Mission with the Science Activity Planner."

12. Schön, *Educating the Reflective Practitioner*.

13. Michael Meltzer, "The Cassini History Project," NASA 2009 History Program Review and Training, NASA/Ames, April 30.

14. Bell, "Photographing Mars," 16.

15. Oran W. Nicks, *Far Travelers: The Exploring Machines* (Washington, DC: National Aeronautics and Space Administration Special Publication-480, 1985), 246.

16. Nicks, *Far Travelers*, 245.

17. Catherine Wilson, "Vicariousness and Authenticity," in Goldberg, *The Robot in the Garden*, 65–88. Mediated agency is defined as "action on a real object at a distance" (79).

18. Antonella Carassa, Francesca Morganti, and Maurizio Tirassa, "Movement, Action, and Situation: Presence in Virtual Environments," in *Proceedings of the 7th Annual International Workshop on Presence* (*Presence 2004*), ed. M. Alcañiz Raya and B. Rey Solaz (Valencia, Spain, October 13–15, 2004), 7–12; quotation cited 7.

19. Launius and McCurdy, *Robots in Space*, 200.

20. Matthew Lombard and Teresa Ditton, "At the Heart of It All: The Concept of Presence," *Journal of Computer-Mediated Communication* 3, no. 2 (1997), http://jcmc.indiana.edu/vol3/issue2/lombard.html (accessed June 22, 2009).

21. Ibid.

22. Wilson, "Vicariousness and Authenticity," 79.

23. Launius and McCurdy, *Robots in Space*, 20; see also Chaikin, "The Other Moon Landings"; NASA GSFC, "Soviet Lunar Missions," http://nssdc.gsfc.nasa.gov/planetary/lunar/lunarussr.html (accessed July 27, 2009); NASA National Space Science Data Center, "Luna 21/Lunokhod 2," http://nssdc.gsfc.nasa.gov/nmc/masterCatalog.do?sc=1973-001A (accessed July 27, 2009).

24. For a report on the Kilauea field test, see Graham Ryder, "Testing a 'Lunar' Marsokhod," *Lunar and Planetary Information Bulletin* 75 (Spring 1995), http://www.lpi.usra.edu/publications/newsletters/lpib/lpib75/bull4.html (accessed March 7, 2008).

25. For an excellent overview of MER driving methods, see Jeffrey J. Biesiadecki, Chris Leger, and Mark W. Maimone, "Tradeoffs Between Directed and Autonomous Driving on the Mars Exploration Rovers," *The International Journal of Robotics Research* 26, no. 1 (2007): 91–104.

26. JPL MER Mission Updates, "Sol 81–82, April 17, 2004: Record-Setting Drive," http://marsrovers.jpl.nasa.gov/mission/status_opportunityAll_2004.html#sol81 (accessed July 11, 2008).

27. Wikipedia, "Mars Exploration Rover," http://en.wikipedia.org/wiki/Mars_Exploration_Rover (accessed July 22, 2011); Launius and McCurdy, *Robots in Space*, 155.

28. Steve Chien, Rebecca Castano, Benjamin Bornstein, Alex Fukunaga, Andres Castano, Jeffrey Biesiadecki, Ron Greeley, Patrick Whelley, and Mark Lemmon, "Results from Automated Cloud and Dust Devil Detection Onboard the MER Rovers," *International Symposium on Artificial Intelligence, Robotics and Automation for Space (i-SAIRAS 2008),* Universal City, CA, February 2008.

29. JPL MER Press Releases, "March 23, 2010: NASA Mars Rover Getting Smarter as It Gets Older," http://marsrovers.jpl.nasa.gov/newsroom/pressreleases/20100323a.html (accessed March 27, 2010).

30. Tara Estlin, Rebecca Castano, Benjamin Bornstein, Daniel Gaines, Robert C. Anderson, Charles de Granville, David Thompson, Michael Burl, Michele Judd, and Steve Chien, "Automated Targeting for the MER Rovers," *Proceedings of the Space Mission Challenges for Information Technology Conference (SMC-IT 2009)*, Pasadena, CA, July 2009, 267–273.

31. For an excellent review of MER's capabilities, see Max Bajracharya, Mark Maimone, and Daniel Helmick, "Autonomy for Mars Rovers: Past, Present, and Future," *Computer* 41, no. 12 (December 2008): 44–50.

32. JPL MER Mission Updates, "Sol 360–366, February 04, 2005: Poking Around on the Plains," http://marsrovers.jpl.nasa.gov/mission/status_opportunityAll_2005.html#sol360 (accessed January 24, 2008).

33. JPL MER Mission, "Press Release Images: Opportunity, 2-Jan-2008, D-Star Panorama by Opportunity," http://marsrovers.jpl.nasa.gov/gallery/press/opportunity/20080102a.html (accessed January 28, 2008).

34. For a related analysis providing more details about the engineers' perspective of the rover as a co-worker and "interlocutor," see Mirmalek, "Solar Discrepancies." For a thorough philosophical analysis of people's mentality in using automated tools, see Bruno Latour, *Pandora's Hope: Essays on the Reality of Science Studies* (Cambridge, MA: Harvard University Press, 1999). In Latour's terms, in a "collective of humans and nonhumans" black-box tools mediate our actions through *delegation* and hence become "agents"—which are subsequently conceived as *autonomous* partners in a form of "fetishism" (174, 187, 273).

35. Nicks, *Far Travelers*, 139–140, emphasis added. See also Wikipedia, "Surveyor Program," http://en.wikipedia.org/wiki/Surveyor_program (accessed January 22, 2008).

36. Surveyor Program, *Surveyor: Program Results* (Washington, DC: National Aeronautics and Space Administration Special Publication-184, 1969), xvi, 32.

37. Interview with Carol Stoker, 14 February 2001. For details about Viz, see Carol Stoker, Eric Zbinden, Theodore T. Blackmon, Bob Kanefsky, Joel Hagen, Charles Neveu, Daryl Rasmussen, Kurt Schwehr, and Michael Sims, "Analyzing Pathfinder Data Using Virtual Reality and Super-resolved Imaging," *Journal of Geophysical Research—Planets* 104, no. E4 (1999): 8889–8906. For a probing discussion of the different visualization tools used for the Phoenix mission, see Vertesi, "Seeing Like a Rover: Images in Interaction on the Mars Exploration Mission," 363–365.

38. For a related analysis on how team members must gain fluency in interpreting visualizations such as the distorted (fish-eye) Hazcam images, see Janet A. Vertesi, "'Seeing Like a Rover': Embodied Experience on the Mars Exploration Rover Mission," 2523–2532.

39. Mark W. Maimone, Chris Leger, and Jeffrey J. Biesiadecki, "Overview of the Mars Exploration Rovers' Autonomous Mobility and Vision Capabilities," *IEEE International Conference on Robotics and Automation (ICRA) Space Robotics Workshop* (Rome, April 14, 2007).

40. Pyne, *Voyager: Seeking Newer Worlds in the Third Great Age of Discovery,* xviii–xix.

41. Nicks, *Far Travelers*, 245.

42. Squyres, *Roving Mars*, 75–92.

43. For example, see Squyres et al., "Athena Mars Rover Science Investigation."

44. Squyres, *Roving Mars*, 325.

45. JPL MER Press Releases, "June 8, 2003: Girl with Dreams Names Mars Rovers 'Spirit' and 'Opportunity,'" http://marsprogram.jpl.nasa.gov/mer/newsroom/pressreleases/20030608a.html (accessed August 8, 2011).

46. JPL, "NASA Names First Rover to Explore the Surface of Mars," http://mars.jpl.nasa.gov/MPF/rover/name.html (accessed August 8, 2011).

47. JPL, Mars Science Laboratory, "Spacecraft: Surface Operations Configuration: Rover," http://mars.jpl.nasa.gov/msl/mission/rover.html (accessed November 23, 2011).

48. JPL Cassini News Releases, http://saturn.jpl.nasa.gov/news/newsreleases/ (accessed August 8, 2011).

49. JPL MER Press Releases, "May 25, 2011: NASA Spirit Rover Completes Mission on Mars," http://marsrovers.jpl.nasa.gov/newsroom/pressreleases/20110525a.html (accessed July 22, 2011).

50. The definition and example refer to figures of speech sometimes called more specifically *synecdoche*, a kind of metonymy. Thus "the White House" could be said to refer to the US president and staff, who are not part of the White House building but are associated with it—an example of metonymy figure of speech that is not a synecdoche. Note the contrast with *metaphor*, such as in calling Pancam "the rover's eyes." See "Metonymy," http://en.wikipedia.org/wiki/Metonymy (accessed January 8, 2007).

51. JPL MER Mission Updates, "Sol 933–942, August 25, 2006: Spirit Continues Mid-Winter Studies of Martian Rocks and Soil," http://marsrovers.jpl.nasa.gov/mission/status_spiritAll_2006.html#sol933 (accessed August 25, 2006).

52. JPL MER Mission Updates, "Sols 2063–2068, November 12–17, 2009: 'Marquette' Study Begins," http://marsrovers.jpl.nasa.gov/mission/status_opportunityAll_2009.html (accessed July 28, 2011). Mention of "the team" in JPL status reports peaks in 2005 [85 mentions for Spirit; 239 for Opportunity]. This wording fits the variety and intensity of the investigation [figure 2.4] that year and drops steadily thereafter to only seven mentions altogether in 2010.

53. For a related analysis of the personification of the rovers, see Mirmalek, "Solar Discrepancies." Although Mirmalek and I worked together on the MER project in 2004 at JPL, we formulated our interpretations and presentations and wrote independently in the years that followed. See also Vertesi, "Seeing Like a Rover: Images in Interaction on the Mars Exploration Mission," for further discussion of how the personification of the rover unifies the team.

54. See Clancey, "Roles for Agent Assistants in Field Science: Personal Projects and Collaboration," for further explanation of personal projects and human consciousness.

55. Squyres, quoted by Carberry, "Interview with Dr. Steve Squyres," 13.

56. Clancey, "Situated Robots," in *Situated Cognition.*

57. Carassa, Morganti, and Tirassa, "Movement, Action, and Situation," 11.

58. Ibid.

59. For a thorough explanation of how human mental capabilities differ from model-based machines, particularly the nature of consciousness, see Clancey, *Situated Cognition* and *Conceptual Coordination*.

60. Squyres, January 2008, quoted by Chaikin, *A Passion for Mars*, 263.

61. Andrea Thompson, "Steve Squyres: Robot Guy Says Humans Should Go to Mars," *Space.com*, July 15, 2009, http://www.space.com/6972-steve-squyres-robot-guy-humans-mars.html (accessed July 22, 2009).

62. JPL MER Mission Press Release, "Meteorite Found on Mars Yields Clues about Planet's Past," August 10, 2009, http://marsrovers.jpl.nasa.gov/newsroom/pressreleases/20090810a.html (accessed August 26, 2009).

63. David Des Marais, email message to author, May 11, 2011.

64. Michael Sims, email message to author, May 11, 2011.

65. JPL MER Mission Updates, "Sols 2601–2607, May 19–25, 2011: Opportunity Spies Outcrop Ahead," http://marsrovers.jpl.nasa.gov/mission/status_opportunityAll_2011.html#sol2601 (accessed May 11, 2011).

66. Launius and McCurdy, *Robots in Space*, 21.

67. Rudwick, "Geological Travel and Theoretical Innovation," 154.

68. JPL MER Mission Updates, "Sol 1445–1449, January 30, 2008: Spirit Takes Steps to Conserve Energy During Martian Winter," http://marsrovers.jpl.nasa.gov/mission/status_spiritAll_2008.html (accessed February 19, 2008).

69. Bell, "Photographing Mars," 16.

7   The Communal Scientist

1. Unless noted otherwise, all quotations from Cabrol, Carr, Des Marais, Matijevic, Rice, Sims, Squyres, Trebi-Ollennu, and Yingst are from the interviews cited chapter 5, note 3.

2. Doug Ellison, "Steve Squyres Q'n'A," Unmannedspaceflight.com, 2005, http://www.unmannedspaceflight.com/index.php?s=6c5277a5fa9fee6ce20e17c7fb6a9419&showtopic=1683 (accessed January 15, 2008).

3. Vertesi, "Seeing Like a Rover: Images in Interaction on the Mars Exploration Mission," 181.

4. Michael Meltzer, *Mission to Jupiter: A History of the Galileo Project* (Washington, DC: National Aeronautics and Space Administration Special Publication-4231, 2007), 122, 131–132, 142.

5. Vertesi, "Seeing Like a Rover: Images in Interaction on the Mars Exploration Mission," 171.

6. Michael Meltzer, personal communication, April 30, 2009.

7. Very roughly, given the $400 million cost per rover and assuming an average of eight observations a day per rover over five years, the amortized cost of each observation would be over $25,000. A single "observation" may involve many images or automated instrument control, so direct comparison of productivity and costs to scientists on an earth expedition is difficult.

8. Squyres, *Roving Mars*, 97.

9. JPL MER Mission Press Releases, "May 25, 2011: NASA Spirit Rover Completes Mission on Mars," http://marsrovers.jpl.nasa.gov/newsroom/pressreleases/20110525a.html (accessed July 22, 2011).

10. Squyres, *Roving Mars*, 315.

11. Squyres et al., "Athena Mars Rover Science Investigation."

12. Interview with Matt Golombek, March 22, 2001.

13. Interview with Jim Bell, February 22, 2001, by telephone, with Jay Trimble and Roxana Wales; field notes provided by Jay Trimble.

14. For details about the original design of intent frames and the naming scheme, see Valerie Shalin, "Final Report for NCC2-1411: Participant Observation of MER Design and Development: Implications for the Relationship between Tools and Scientific Reasoning" (Dayton, OH: Wright State University, November 2004); Valerie Shalin, Roxana Wales, and Deborah Bass, "Communicating Intent for Planning and Scheduling Tasks," *Proceedings of HCI-International 2005 Conference* (Mahwah, NJ: Lawrence Erlbaum Associates, 2005); and Wales, Shalin, and Bass, "Requesting Distant Robotic Action."

15. Kohler, *Landscapes and Labscapes*, 194.

16. For example, see the research of Nick Tosca (MER summer 2004 student intern), "Planetary and Terrestrial Sedimentary Research," http://www.ic.sunysb.edu/project/mersb/people_nick.htm (accessed June 25, 2009), and a project in Scott McLennan's group, http://www.ic.sunysb.edu/project/mersb/people_scott.htm (accessed June 25, 2009).

17. Phil Christensen, "Fusing Science and Engineering," School of Earth and Space Exploration, Arizona State University, http://themis.mars.asu.edu/christensen (accessed June 25, 2009).

18. M. Darby Dyar, Mount Holyoke Faculty Profile, http://www.mtholyoke.edu/acad/faculty-profiles/darby_dyar.html (accessed July 25, 2009).

19. M. Darby Dyar, E. Murad, E. C. Sklute, J. L. Bishop, and A. C. Muirhead, "Mössbauer and Reflectance Spectroscopy of Iron Oxide Mixtures," *40th Lunar and Planetary Science Conference*, abstract no. 2209 (2009); R. Aileen Yingst, M. E. Schmidt, R. C. F. Lentz, M. J. Christman, and R. Behnke, "Understanding Mars at the Microscale by Imaging Terrestrial Analogs: The Handlens Atlas," *38th Lunar and Planetary Science Conference*, abstract no. 1130 (2007).

20. Kohler, *Landscapes and Labscapes*, 308.

21. Squyres, *Roving Mars*, 328.

22. Joseph N. Tatarewicz, "The Hubble Space Telescope Servicing Mission," in Pamela Mack, *From Engineering Science to Big Science* (Washington, DC: National Aeronautics and Space Administration Special Publications-4219, 1998), 374.

23. Mishkin, *Sojourner*, 235.

24. Squyres, *Roving Mars*, 329.

25. PDS Geosciences Node, Washington University, "MER Analyst's Notebook User's Guide" (2006).

26. Tunstel et al., "FIDO Rover Field Trials as Rehearsal for the NASA 2003 Mars Exploration Rovers Mission."

27. The HCC team consisted of NASA Ames and university psychologists and social and computer scientists and Deborah Bass, the MER Science Operations Systems Engineer and later the MER Deputy Science Team Chief.

28. Roxana Wales and Alonso Vera, "Failable [sic] Metrics: Assessing the Outcome of Deploying Collaborative Technology into the MER Mission," Presentation to Athena Science Team, August 27, 2003.

29. Roxana Wales, personal communication, quoted from field notes "FIDO Mars Yard Test, Sol 2" (March 29, 2001).

30. Roxana Wales, personal communication, from field notes "FIDO Mars Yard Test, Sol 4" (March 30, 2001).

31. Roxana Wales, Valerie Shalin, and Deborah Bass, "Lessons Learned from the January '03 Science Team Trainings," Presentation to Athena Science Team, March 13, 2003.

32. Roxana Wales and Valerie Shalin, "Lessons Learned: Issues in Identification/Naming, SOWG Process and Intent Transmission Thread Test H," Presentation to Athena Science Team, June 22, 2003.

33. Valerie Shalin and Roxana Wales, "Recommendations for an Efficient Search Procedure When Developing a Science Activity Plan," Presentation to Athena Science Team, June 22, 2003.

34. Wales et al., "Requesting Distant Robotic Action," 96.

35. PDS Geosciences Node, "MER Analyst's Notebook User's Guide," 94.

36. JPL, "Object Naming," Flight Team Procedure for Science Operations Working Group no. SOWG-TP-02 (January 17, 2004).

37. Wales et al., "Requesting Distant Robotic Action," 84.

38. Roxana Wales, personal communication, quoted from field notes "FIDO Mars Yard Test, Sol 2" (March 29, 2001).

39. Paul G. Backes, Jeffrey S. Norris, Mark W. Powell, Marsette A. Vona, Robert Steinke, and Justin Wick, "The Science Activity Planner for the Mars Exploration Rover Mission: FIDO Field Test Results," *Proceedings IEEE Aerospace Conference* 8 (March 8–15, 2003), 8-3525–8-3539; Vona et al., "Challenges in 3D Visualization for Mars Exploration Rover Mission Science Planning."

40. Wales et al., "Requesting Distant Robotic Action," 97.

41. JPL, "Tactical Science Activity Plan Generation and Delivery," Flight Team Procedure for Science Operations Working Group no. SOWG-TP-04 (January 17, 2004).

42. John L. Bresina, Ari K. Jónsson, Paul H. Morris, and Kanna Rajan, "Activity Planning for the Mars Exploration Rovers," in *Proceedings of the Fifteenth International Conference on Automated Planning and Scheduling (ICAPS 2005)*, ed. Susanne Biundo, Karen Myers, and Kanna Rajan (Menlo Park, CA: AAAI Press, 2005), 40–49.

8   The Scientist Engineers

1. Squyres, *Roving Mars,* 75. See *Roving Mars* for the long history of the MER rovers and the relationships between scientists, instrument makers, and NASA. More generally, for the history of science on instruments, see, for example, Peter Louis Galison, *Image and Logic* (Chicago: University of Chicago Press, 1997).

2. H. Russell Bernard and Peter. D. Killworth, "Scientists and Crew: A Case Study in Communications at Sea," *Maritime Studies and Management* 2 (1974): 112–125. See also William J. Clancey, "Field Science Ethnography: Methods for Systematic Observation on an Expedition," *Field Methods* 13, no. 3 (August 2001): 223–243.

3. Squyres, *Roving Mars*, 98.

4. For an SOWG room layout diagram, see De Jong et al., "The Benefits of Virtual Presence in Space (VPS) to Deep Space Missions."

5. Ezell and Ezell, *On Mars*, 338.

6. Unless noted otherwise, all quotations from Cabrol, Carr, Des Marais, Matijevic, Rice, Sims, Squyres, Trebi-Ollennu, and Yingst are from the interviews cited in chapter 5, note 3.

7. Susan Leigh Star and James R. Griesemer, "Institutional Ecology, 'Translations,' and Boundary Objects: Amateurs and Professionals in Berkeley's Museum of Vertebrate Zoology, 1907–1939," *Social Studies of Science* 19 (1989): 387–420; quotation cited 409.

8. Michael Sims, email message to author, July 8, 2008. JPL MER Mission Updates, "Sol 1567–1573, May 30–June 5, 2008: Not Quite Hibernation," http://marsrovers.jpl.nasa.gov/mission/status_spiritAll_2008.html#sol1567 (accessed July 11, 2008).

9. Nathalie Cabrol, Edmund A. Grin, and David Fike, "Gusev Crater: A Landing Site for MER," *38th Lunar and Planetary Science Conference*, abstract no. 1142 (2002).

10. JPL, "Deep Space 1," http://nmp.nasa.gov/ds1/ (accessed May 27, 2011).

11. Squyres, *Roving Mars*, 104.

12. JPL MER Mission Updates, "Sol 156–158, June 16, 2004: Spirit Reaches the 'Columbia Hills'!" http://marsrovers.jpl.nasa.gov/mission/status_spiritAll_2004.html#sol156 (accessed January 16, 2008); emphasis added.

13. Steve W. Squyres et al., "In Situ Evidence for an Ancient Aqueous Environment at Meridiani Planum, Mars," *Science* 306 (December 3, 2004): 1709–1714.

14. For detailed analyses of the use of computer simulation compared to physical experiments, see Turkle, *Simulation and Its Discontents*.

15. Charlotte Linde, "Learning from the Mars Rover Mission: Scientific Discovery, Learning, and Memory," *Journal of Knowledge Management, Special Issue on Space Knowledge Management* 10, no. 2 (2006), 90–102. Linde describes the varieties of learning that occurred in preparations and during the mission regarding instruments, models, and planning.

9 The Personal Scientist

1. Unless noted otherwise, all quotations from Cabrol, Carr, Des Marais, Matijevic, Rice, Sims, Squyres, Trebi-Ollennu, and Yingst are from the interviews cited in chapter 5, note 3.

2. Rob Haggart, "Tim Hetherington's Last Interview," *Outside* (July 2011), 96–97.

3. Rudwick, "Geological Travel and Theoretical Innovation," 148.

4. Ibid., 157.

5. Ibid., 150.

6. Ibid., 148.

7. Ibid., 149.

8. Bell, "Photographing Mars," 17.

9. Rudwick, "Geological Travel and Theoretical Innovation," 149.

10. Thomas J. Campanella, "Eden by Wire: Webcameras and the Telepresent Landscape," in Goldberg, *The Robot in the Garden*, 22–46.

11. Bell, "Photographing Mars," 16–17.

12. JPL, "Mars Public Engagement: FIDO Field Test Training Day," presentation, August 9, 2002.

13. Squyres, *Roving Mars*, 246. The IMAX video was produced by Disney Studios.

14. Nicks, *Far Travelers*, 246.

15. This sequence occurred on Spirit, sols 469–470.

16. Bell, "Photographing Mars," 2, 15.

17. Jim Bell, *Postcards from Mars* (New York: Dutton [Penguin Group], 2006), 1–3.

18. Wikipedia, "William Hodges," http://en.wikipedia.org/wiki/William_Hodges (accessed August 17, 2009).

19. National Maritime Museum, "William Hodges 1774–1797: The Art of Exploration," http://www.nmm.ac.uk/upload/package/30/home.php (accessed March 31, 2008).

20. Nicolson, "Historical Introduction," xi.

21. Ibid., xix–xx.

22. Merriam-Webster, "Romantic," *Merriam-Webster Unabridged*, http://www.merriam-webster.com/dictionary/romantic http://unabridged.merriam-webster.com/ (accessed July 6, 2009).

23. Encyclopedia Britannica, Inc., "Romanticism," *British Concise Encyclopedia*, http://www.answers.com/library/Britannica%20Concise%20Encyclopedia-cid-69942 (accessed July 7, 2009).

24. Dyson, *The Scientist as Rebel*, 16.

25. Ibid., 17.

26. Chaikin, *A Passion for Mars*, 104.

27. Rudwick, "Geological Travel and Theoretical Innovation," 157.

28. Kohler, *Landscapes and Labscapes*, 10, citing Rudwick, 1996.

29. Rudwick, "Geological Travel and Theoretical Innovation," 145.

30. Ibid., 151.

31. Rudwick, "Geological Travel and Theoretical Innovation," 156.

32. A. J. S. Rayl, "Mars Exploration Rovers Update: Opportunity Closes in on Endeavour as Team Bids Farewell to Spirit," *Planetary News*, July 31, 2011, http://planetary.org/news/2011/0731_Mars_Exploration_Rovers_Update.html (accessed 9 August 2011).

33. Alain De Botton, *The Art of Travel* (New York: Pantheon, 2002), 183.

34. Ibid., 135.

35. In an independent analysis, Vertesi in "Seeing Like a Rover: Images in Interaction on the Mars Exploration Mission" discusses "the aesthetic of the martian picturesque" in considerable detail, including the work of creating the images, the relation to "American frontier photography," and the effect on the public.

36. Squyres, *Roving Mars*, 251.

37. Interview, May 15, 2001, quoted in *Mission to Jupiter*, 291.

38. Squyres, *Roving Mars*, 377.

39. For a detailed story of the "memorial" celebration of Spirit's mission, see Rayl, "Mars Exploration Rovers Update: Opportunity Closes in on Endeavour as Team Bids Farewell to Spirit."

10   The Future of Planetary Surface Exploration

1. Fred E. Emery and Eric L. Trist, "Socio-Technical Systems," in *Management Sciences, Models and Techniques*, ed. C. W. Churchman (London: Pergamon, 1960).

2. Pyne, "Seeking Newer Worlds: An Historical Context."

3. For a wide-ranging discussion of the development of experimentation and its relation to institutions, see Fara, *Science: A Four Thousand Year History*.

4. The Haughton-Mars Project is an example of a modern expedition focusing on Mars analog science and exploration. See "Haughton-Mars Project," http://www.marsonearth.org (accessed January 24, 2008).

5. British Antarctic Survey, "Halley Research Station," http://www.antarctica.ac.uk/living_and_working/research_stations/halley/ (accessed January 28, 2008).

6. For a discussion of how "big technology" at NASA is a kind of "big science," see Mack, *From Engineering Science to Big Science*.

7. Robert D. Ballard, A.M. McCann, D. Yoerger, L. Whitcomb, D. Mindell, J. Oleson, H. Singh, B. Foley, J. Adams, and D. Picheota, "The Discovery of Ancient History in the Deep Sea Using Advanced Deep Submergence Technology," *Deep-Sea Research I* 47, no. 9 (September 2000), 1591–1620.

8. Mindell and Brigham, "New Archaeological Uses of Autonomous Undersea Vehicles"; Stefan Helmreich, "Intimate Sensing," in Sherry Turkle, *Simulation and Its Discontents*, 129–150; Stefan Helmreich, "An Anthropologist Underwater: Immersive Soundscapes, Submarine Cyborgs, and Transductive Ethnography," *American Ethnologist* 34, no. 4 (2007): 621–641; Brian Bingham, Brendan Foley, Hanumant Singh, Richard Camilli, Katerina Delaporta, Ryan Eustice, Angelos Mallios, David Mindell, Christopher Roman, and Dimitris Sakellariou, "Robotic Tools for Deep Water Archaeology: Surveying an Ancient Shipwreck with an Autonomous Underwater Vehicle," *Journal of Field Robotics*, Special Issue on State of the Art in Maritime Autonomous Surface and Underwater Vehicles, 27, no. 6 (November–December 2010): 702–717.

9. See Donald A. Beattie, *Taking Science to the Moon: Lunar Experiments and the Apollo Program* (Baltimore: John Hopkins University Press, 2001), 95–96, 125–145; quotation cited xiv.

10. Dyson, *The Scientist as Rebel*, 207.

11. Ibid., 208.

12. For a related discussion about the interplay of field science and engineering, see David Mindell and Katherine L. Croff, "Deep Water, Archaeology, and Technology Development," *MTS Journal* 36, no. 3 (2002): 13–20.

13. Squyres, quoted in Carberry, "Interview with Dr. Steve Squyres," 13.

14. Thomas Kuhn, *Structure of Scientific Revolutions* (Chicago: University of Chicago, 1962).

15. Rayl, "Mars Exploration Rovers Update: Opportunity Closes in on Endeavour as Team Bids Farewell to Spirit."

16. Squyres, *Roving Mars*, 244, 251.

17. Unless noted otherwise, all quotations from Cabrol, Carr, Des Marais, Matijevic, Rice, Sims, Squyres, Trebi-Ollennu, and Yingst are from the interviews cited in chapter 5, note 3.

18. Quoted by Vertesi, "'Seeing Like a Rover': Embodied Experience on the Mars Exploration Rover Mission," 2526.

19. Clancey, "Clear Speaking about Machines."

20. Jason M. Fox and Michael McCurdy, "Activity Planning for the Phoenix Mars Lander Mission," *Proceedings IEEE Aerospace Conference* (March 3–10, 2007), 1–13.

21. Wick et al., "Distributed Operations for the Mars Exploration Rover Mission with the Science Activity Planner."

22. For example, the EPOXI mission was retargeted to a second comet after Deep Impact's encounter with the comet 9P/Tempel.

23. Joseph N. Tatarewicz, "The Hubble Space Telescope Servicing Mission," 374.

24. Phoenix's wet chemistry lab Microscopy, Electrochemistry, and Conductivity Analyzer (MECA) could process four samples; the Thermal and Evolved Gas Analyzer (TEGA) had slots for eight samples. NASA Phoenix Mars Lander, http://www.nasa.gov/mission_pages/phoenix/main/index.html (accessed July 4, 2008). Regarding Scout missions, see NASA Mars Exploration Program, "Beyond 2005," http://mars.jpl.nasa.gov/missions/future/2005-plus.html (accessed January 29, 2008).

25. "Human and Robots in Exploration," workshop sponsored by the Search for Extraterrestrial Intelligence (SETI) and the Planetary Society, Stanford University, February 12–13, 2008.

26. Clancey, *Situated Cognition*. For a more elaborate discussion of the use of robots in exploration—particularly contrasting coordination, cooperation, and collaboration—see Clancey, "Roles for Agent Assistants in Field Science."

27. JPL Special Review Board, "Report on the Loss of the Mars Polar Lander and Deep Space 2 Missions," JPL D-18709 (March 22, 2000).

28. Jim Brown, Robots.Net commentary on "No Autonomous Mining Robots, Says Minsky," April 18, 2005, http://robots.net/article/1477.html (accessed July 25, 2011).

29. *Insurance Journal*, "Federal Figures Show 32 Coal Miners Killed in 2007," http://www.insurancejournal.com/news/national/2008/01/03/86063.htm (accessed January 31, 2008). Regarding China, see "Disasters," *Encyclopedia Britannica 2007 Book of the Year* (Chicago: Encyclopedia Britannica, 2007), 58.

30. John Callas, email message to author, May 13, 2011.

31. Andrew Mishkin and Sharon Laubach, "From Prime to Extended Mission: Evolution of the MER Tactical Uplink Process," JPL, California Institute of Technology, 2006, http://trs-new.jpl.nasa.gov/dspace/bitstream/2014/39916/1/06-1526.pdf (accessed May 13, 2011).

32. John Callas, email message to author, May 13, 2011.

33. Andy Mishkin, email message to author, May 16, 2011.

34. The tool is called AutoPUL (referring to payload uplink leads). It automates nearly all of the nominal operation of PULs for the MI and engineering cameras (Hazcams and Navcams) and much of the off-nominal activities. It does not automate unusual sequences or alleviate the approval overview process: "Seven PULs are responsible for generating commands for the following sol that must be uploaded to the instruments and the Pancams on the rovers at the end of each day." Thomas Oberst, "Restless Rovers Demand Long Hours from CU's 'Martians,'" *Cornell Chronicle*, http://www.news.cornell.edu/Chronicle//04/11.11.04/MarsLab.html (accessed January 31, 2008).

35. Andy Mishkin, email message to author, May 16, 2011.

36. For a study of how the team became more efficient, see Irene Tollinger, Christian D. Schunn, and Alonso Vera, "What Changes When a Large Team Becomes More Expert?," in *Proceedings of the 28th Annual Conference of the Cognitive Science Society* (Mahwah, NJ: Lawrence Erlbaum Associates, 2006), 840–845.

37. For an excellent, detailed summary of the changing requirements and methods of MER operations, see Anthony Barrett, Deborah Bass, Sharon Laubach, and Andrew Mishkin,

"A Retrospective Snapshot of the Planning Processes in MER Operations After 5 Years," Space Mission Challenges for Information Technology (SMC-IT), 2009.

38. Ashitey Trebi-Ollennu and Antonio Diaz-Calderon, "Planned Activity Complexity Evaluation (PACE): Applied to Mars Exploration Rovers Surface Activities," JPL Publication 07-4, 14, 19-20.

39. Steve Squyres, email message to author, June 6, 2011.

40. A. J. S. Rayl, "Mars Exploration Rovers: Spirit Slides into Home Plate as Opportunity Finishes Work at Erebus," The Planetary Society, February 28, 2006, http://www.planetary.org/news/2006/0228_Mars_Exploration_Rovers_Update_Spirit.html (accessed March 11, 2008).

41. Michael Sims, email message to author, March 4, 2008.

42. Steve Squyres, email message to author, March 5, 2008.

43. David Des Marais, email message to author, March 5, 2008.

44. JPL MER Mission, Press Releases, "NASA Spacecraft Provides Travel Tips for Mars Rover," December 16, 2010, http://marsrovers.jpl.nasa.gov/newsroom/pressreleases/20101216a.html (accessed July 29, 2011).

45. The Planetary Society, "Mars Exploration Rovers Update," April 30, 2007, http://www.planetary.org/news/2007/0430_Mars_Exploration_Rovers_Update_Spirit.html (accessed January 28, 2008).

46. Michael Sims, email message to author, March 6, 2008.

47. Compare with direct manipulation of distant objects in "virtual reality," Carassa et al., "Movement, Action, and Situation," 12.

48. Richard A. Kerr, "Phoenix Rose Again, But Not All Worked Out as Planned," *Science* 323 (February 11, 2009): 872–873.

49. National Research Council, "Assessment of Planetary Protection Requirements for Mars Sample Return Missions," May 2009; executive summary available at http://www.nap.edu/catalog.php?record_id=12576 (accessed May 14, 2009).

50. A. J. S. Rayl, "Mars Exploration Rovers Update: Spirit Gains a Little Power, Opportunity Loses a Little Steam," *Planetary News*, February 28, 2009, http://planetary.org/news/2009/0228_Mars_Exploration_Rovers_Update_Spirit.html (accessed July 26, 2011).

51. Liam Pedersen, David E. Smith, Matthew Deans, Randy Sargent, Clay Kunz, David Lees, and Srikanth Rajagopalan, "Mission Planning and Targeting for Autonomous Instrument Placement," *Aerospace Conference*, abstract no. 2.0103 (Big Sky, MT, March 3–10, 2007), http://www.aeroconf.org/2007_web/2005%20Digest%20rev%209%20plus%20page%201.pdf (accessed September 19, 2009).

52. Office of the Inspector General, "NASA's Management of the Mars Science Laboratory Project," Report No. 1G-11–019, June 8, 2011.

53. Daniel Lester and Harley Thronson, "Human Space Exploration and Human Spaceflight: Latency and the Cognitive Scale of the Universe," *Space Policy* 27 (2011) 89–93; Daniel Lester and Harley Thronson, "Low-Latency Lunar Surface Telerobotics from Earth-Moon Libration Points," *AIAA SPACE 2011 Conference and Exposition,* AIAA-2011-7341 (Long Beach, CA, September 27–29, 2011). See also George R. Schmidt, Geoffrey A. Landis, Steven R. Oleson, Stanley K. Borowski, and Michael J. Krasowski, "HERRO: A Science-Oriented Strategy for

Crewed Missions Beyond LEO," *48th AIAA Aerospace Sciences Meeting Including the New Horizons Forum and Aerospace Exposition,* AIAA 2010-629 (Orlando, FL, January 4–7, 2010).

54. Jared Diamond, *Guns, Germs, and Steel: The Fates of Human Societies* (New York: W. W. Norton & Co., 1997), 371–372.

54. "Hot spots" are discovered by scanning the earth twice a day at 1 km/pixel resolution, using the Moderate Resolution Imaging Spectroradiometer (MODIS) Thermal Alert System; see Hawaii Institute of Geophysics and Planetology, "Near-Real-Time Thermal Monitoring of Global Hot-Spots," http://modis.higp.hawaii.edu (accessed January 31, 2008).

Epilogue

1. Squyres, January 2008, quoted by Chaikin, *A Passion for Mars*, 259.

2. George W. Bush, *A Renewed Spirit of Discovery: The President's Vision for U.S. Space Exploration* (Washington, DC: January 2004).

3. Alicia Chang, "NASA to Abandon Trapped Rover Spirit," APNewsBreak, May 24, 2011, http://abcnews.go.com/Technology/wireStory?id=13676921 (accessed May 24, 2011).

4. Meltzer, *Mission to Jupiter*, 300.

5. Ibid., 299.

6. John Baross, "The Evolution of Astrobiology: Searching for Life in the Universe—A New Darwinian Voyage," abstract for public lecture, Mountain View Center for the Performing Arts, Mountain View, CA, November 2, 2009.

7. T. S. Eliot, *Four Quartets* (New York: Harcourt, 1943), quotation from "Little Gidding," 59.

8. Casani interview, May 29, 2001, quoted by Meltzer, *Mission to Jupiter*, 287.

9. Dayton Duncan and Ken Burns, *The National Parks: America's Best Idea* (New York: Knopf, 2009).

10. Transcribed from video recorded by author, July 28, 2000.

11. Ibid.

12. Meltzer, *Mission to Jupiter*, 300.

13. Joseph Campbell, *The Hero with a Thousand Faces* (Princeton, NJ: Princeton University Press, 1949). See also Stuart Voytilla, *Myth and the Movies: Discovering the Mythic Structure of 50 Unforgettable Films* (Studio City, CA: Michael Wiese Productions, 1999). See also Stephen J. Pyne, *Voyager: Seeking Newer Worlds in the Third Great Age of Discovery*, 356–364, for a historical, literary discussion of how the mythical narrative of the hero relates to Western exploration and the Voyager mission in particular. Befitting the effect of a cultural archetype, my presentation was developed independently.

14. Michael Griffin, "Statement by NASA Administrator Griffin Before the House Appropriations Subcommittee on Commerce, Justice, Science, and Related Agencies," February 13, 2008, http://www.nasa.gov/pdf/211844main_House_Science_Committee_Oral_13_Fe%2008.pdf (accessed March 6, 2008).

15. Public comments during "Discussion Forum on President's Commission for Implementation of U.S. Space Policy (Moon, Mars, and Beyond)," *Lunar & Planetary Science Conference*, Houston, March 15, 2004.

16. Walter Kaufmann, "Freud and His Poetic Science," in *Discovering the Mind: III. Freud Versus Adler and Jung* (New York: McGraw Hill, 1980), 104, 109. Kaufmann argues that Freud's psychoanalytic focus on myths and literary allusions of experience, as in the analysis of dreams, constitutes a "poetic science of the mind."

Bibliographic Essay

This bibliographic essay provides the reader with a summary of the most important references cited in this work, as well as additional reading to learn more about the technical and methodological aspects of the MER mission and the ethnographic study of doing field science with a mobile, robotic laboratory.

The Mission: Mars Exploration Rovers

Basic information about the Mars Exploration Rovers was posted throughout the mission at the JPL website: http://marsrovers.jpl.nasa.gov/home/index.html. This site includes regular "updates" ordered by rover and date, with annual sequences on archive pages. The dates indicate when the entry was posted on the Web, and may be later than the indicated sol (day on Mars). Periodic "press releases" provide summary accounts of important findings or milestones in each rover's progress. Images are also collected in archival form. Many of the images in this book come from these pages; all of the facts presented here were provided by or crosschecked with the rover updates.

The MER missions have been described by Steve Squyres, the principal investigator, in journal form. His book provides a useful cross-check and comparison for the interviews and ethnographic observations, particularly during the February 2004 El Capitan campaign. As the title indicates, Squyres emphasizes the development of the rovers and instruments:

Squyres, Steve. *Roving Mars: Spirit, Opportunity, and the Exploration of the Red Planet*. New York: Hyperion, 2005.

A program manager's perspective is provided by:

Hubbard, Scott. *Exploring Mars: Chronicles from a Decade of Discovery*. Tucson: University of Arizona Press, 2011.

Two dissertations provide broad introductions while analyzing unique aspects of the scientific work. Mirmalek focuses on the management of time conceptually, in activities, and reified in the documents and tools. Vertesi considers the scientists' experience, particularly the embodied, organizational, and aesthetic aspects of the visual work of creating and interpreting images:

Mirmalek, Zara. "Solar Discrepancies: Mars Exploration and the Curious Problem of Interplanetary Time." PhD diss., Department of Communication, University of California, San Diego, 2008.

Vertesi, Janet A. "Seeing Like a Rover: Images in Interaction on the Mars Exploration Mission." PhD diss., Science and Technology Studies Department, Cornell University, 2009.

The scientific results are detailed in numerous journal and conference publications, with overviews provided in these *Science* articles:

Squyres, Steve W., et al. "The Spirit Rover's Athena Science Investigation at Gusev Crater, Mars." *Science* 305 (5685) (August 6, 2004): 794–799.

Squyres, Steve W., et al. "The Opportunity Rover's Athena Science Investigation at Meridiani Planum, Mars." *Science* 306 (December 3, 2004): 1698–1703.

Squyres, Steve W., et al. "Two Years at Meridiani Planum: Results from the Opportunity Rover." *Science* 313 (September 8, 2006): 1403–1407.

Squyres, Steve W., et al. "Exploration of Victoria Crater by the Mars Rover Opportunity." *Science* 324 (May 22, 2009): 1058–1061.

Bell's book presents improved versions of many MER Pancam images, including double-page spreads; it emphasizes the aesthetics and provides a firsthand story of the expeditions. Carr's revised textbook is a compendium of Mars facts and interpretations, including material about the MER campaigns. Chaikin's personal perspective spans the Mars missions through MER. His presentation is inspirational and positive, partly autobiographical, and emphasizes the individual scientists' and engineers' motivations and experience:

Bell, Jim. *Postcards from Mars*. New York: Dutton (Penguin Group), 2006.

Carr, Michael H. *The Surface of Mars*. New York: Cambridge University Press, 2007.

Chaikin, A. *A Passion for Mars: Intrepid Explorers of the Red Planet*. New York: Abrams, 2008.

A variety of other materials posted on the Internet were used for interpretations of the MER mission; these appear in the chapter notes.

Robotic Systems and Human-Robotic Interaction

The literature on robotic systems and human-robotic interaction is quite large, starting with early work in artificial intelligence in the late 1960s. Research ranges from practical issues of designing mobile sensing systems to philosophical analyses on the nature of knowledge and what it means to be an "agent."

Straddling the topic of robotics and spaceflight, Nicks provides a good starting point with overviews of missions such as Surveyor and reflections on exploration systems that relate human and automated operations. Hubbard provides a more recent summary of robotic methods and perspectives for designing exploration systems; Stoker focuses specifically on Mars exploration methods. Synthesizing critical and practical perspectives, my journal article provides examples of how proactive computer programs ("agents") can be used in surface exploration, with examples from Apollo:

Nicks, Oran W. *Far Travelers: The Exploring Machines.* Washington, DC: National Aeronautics and Space Administration, Special Publication 480, 1985.

Hubbard, G. Scott. "Humans and Robots: Hand in Grip." *Acta Astronaut* 57 (2–8) (2005): 649–660.

Stoker, Carol R., and Carter Emmart. *Strategies for Mars: A Guide to Human Exploration*. San Diego: Univelt, 1996.

Clancey, William J. "Roles for Agent Assistants in Field Science: Personal Projects and Collaboration." *IEEE Transactions on Systems, Man and Cybernetics. Part C, Applications and Reviews* 34 (2) (2004): 125–137.

The modern controversy about "intelligent" machines began in the 1960s, formulated most notably in Dreyfus's philosophical argument that human knowledge and capability cannot be equated with rules and stored descriptions in computer programs. Schön, building on the work of American philosopher John Dewey, clarified that models of the world and how to behave (such as rules and procedures) are not rotely followed, but are *tools* used by people in their interactive behavior as they interpretively "think in action." Influenced by these philosophical analyses, computer scientists specializing on AI provided more technical critiques, showing the inherent limitations of a computer program whose actions are driven by stored models:

Dreyfus, Hubert. *What Computers Can't Do: A Critique of Artificial Reason*. New York: Harper & Row, 1972.

Schön, Donald. *The Reflective Practitioner: How Professionals Think in Action*. New York: Basic Books, 1983.

McDermott, Drew. "Artificial Intelligence Meets Natural Stupidity." *ACM Sigart Newsletter* 57 (April 1976): 4–9.

Winograd, Terry, and Fernando Flores. *Understanding Computers and Cognition: A New Foundation for Design*. Norwood, MA: Ablex, 1986.

Brooks, Rodney A. "Intelligence without Representation." *Artificial Intelligence Journal* 47 (1991): 139–159.

Clancey, William J. *Situated Cognition: On Human Knowledge and Computer Representations*. New York: Cambridge University Press, 1997.

At the time of this writing, nobody knows how the human brain works well enough to replicate in machines the everyday human ability to adaptively coordinate conceptual and perceptual-motor subsystems in real time in ordinary behavior such as speaking, let alone improvisation. One perspective is that the nature of consciousness must be understood if the flexibility and nature of human conceptualization is to be replicated in computer programs. For an analysis of the nature of such a "process memory," see:

Clancey, William J. *Conceptual Coordination: How the Mind Orders Experience in Time*. Mahwah, NJ: Erlbaum, 1999.

Virtual Presence and Agency
The topics of "presence" and "agency" have expanded over the years from relatively abstract philosophical concepts to practical definitions for computer visualization and virtual reality environments. For academic analyses providing useful frameworks, see:

Carassa, Antonella, Francesca Morganti, and Maurizio Tirassa. "Movement, Action, and Situation: Presence in Virtual Environments." In *Proceedings of the 7th Annual International Workshop on Presence*, ed. M. Alcañiz Raya and B. Rey Solaz, 7–12. Valencia, Spain: Universidad Politécnica de València, 2004.

Durlach, Nathaniel I., Thomas B. Sheridan, and Stephen R. Ellis, eds. *Human Machine Interfaces for Teleoperators and Virtual Environments*. NASA CP 100071.

Goldberg, Ken, ed. *The Robot in the Garden: Telerobotics and Telepistemology in the Age of the Internet*. Cambridge, MA: MIT Press, 2000.

Lombard, Matthew, and Teresa Ditton. "At the Heart of It All: The Concept of Presence." *Journal of Computer-Mediated Communication* 3 (2) (1997).

Rudwick, Martin. "Geological Travel and Theoretical Innovation: The Role of `Liminal' Experience." *Social Studies of Science* 26 (1996): 143–159.

Sheridan, Thomas B. "Teleoperation, Telerobotics and Telepresence: A Progress Report." *Control Engineering Practice* 3 (2) (1995): 205–214.

Earlier Space Exploration Missions

Comprehensive publications about NASA's space exploration missions can be found at http://history.nasa.gov/series95.html. Understanding the scientific exploration of the Earth's moon is a good place to begin. The report on Surveyor provides details about each lander's design, operations, and results. (No books about Lunokhod were found; the online sources cited in the notes provide basic facts and photographs.) Beattie's historical review of the scientific work in Apollo describes especially well how the program was formulated. Mindell looks closely at Apollo to understand the work system of people and technology:

Surveyor Program. *Surveyor: Program Results*. Washington, DC: National Aeronautics and Space Administration, Special Publication 184, 1969.

Beattie, Donald A. *Taking Science to the Moon: Lunar Experiments and the Apollo Program*. Baltimore, MD: John Hopkins University Press, 2001.

Mindell, David A. *Digital Apollo: Human and Machine in Spaceflight*. Cambridge, MA: MIT Press, 2008.

Viking's mission operations are particularly well documented:

Corliss, William R. *The Viking Mission to Mars*. Washington, DC: National Aeronautics and Space Administration, Special Publication 334, 1974.

Ezell, Edward C., and Linda N. Ezell. *On Mars: Exploration of the Red Planet: 1958–1978*. Washington, DC: National Aeronautics and Space Administration, Special Publication 4212, 1984.

Lee, B. Gentry. "Mission Operations Strategy for Viking." *Science* 194 (4260) (October 1976): 59–62.

Martin, James S., and A. Thomas Young. "Viking to Mars: Profile of a Space Exploration." *Astronautics & Aeronautics* 14 (November 1976): 22–42, 44–48.

Young, A. Thomas. "Viking Mission Operations." In *Technology Today for Tomorrow: Proceedings of the Twelfth Space Conference,* 6-41–6-48. Cocoa Beach, FL: Canaveral Council of Technical Societies, April 1975.

Andy Mishkin describes the Pathfinder/Sojourner lander from 1997, an experiment whose success strongly motivated the design of MER. In particular, Mishkin describes the science team organization and engineering uplink process that provided a proof of concept for the greatly elaborated organization and process used on MER:

Mishkin, Andy. *Sojourner: An Insider's View of the Mars Pathfinder Mission*. New York: Berkley Publishing Group, 2003.

Meltzer provides a state-of-the-art summary of the instrument packages of the Galileo mission to the Jupiter and its moons. The book ends with interesting stories about the contributions of lead scientists and engineers:

Meltzer, Michael. *Mission to Jupiter: A History of the Galileo Project.* Washington, DC: National Aeronautics and Space Administration, Special Publication 4231, 2007.

At the time of this writing, the Cassini-Huygens mission was still occurring around Saturn and the best source of information was the JPL website: http://saturn.jpl.nasa.gov/index.cfm.

History of Scientific Exploration

Cook's journals provide the primary source of information about his voyages. Forster and Berghof's narrative partly motivated Humboldt's fieldwork; their work exemplifies early scientific exploration. Fara views Humboldt's work as an aspect of the globalization of science:

Edwards, Philip, ed. *The Journals of Captain Cook*. London: Penguin Books, 1999.

Forster, Georg. In *A Voyage Round the World*, ed. Nicholas Thomas and Oliver Berghof. Honolulu: University of Hawaii Press, [1777] 1999.

Humboldt, Alexander von. *Personal Narrative of Journey to the Equinoctial Regions of the New Continent*. London: Penguin Books, [1834], 1995.

Fara, Patricia. *Science: A Four Thousand Year History*. New York: Oxford University Press, 2009.

The relations of discovery, exploration, robotic systems, and human spaceflight have been analyzed and classified by several historians:

Launius, Roger D., and Howard E. McCurdy. *Robots in Space*. Baltimore: Johns Hopkins University Press, 2008.

Pyne, Stephen J. "Seeking Newer Worlds: An Historical Context." In *Critical Issues in the History of Spaceflight*, ed. Steven J. Dick and Roger D. Launius, 7–36. Washington, DC: National Aeronautics and Space Administration, Special Publication 4702, 2006.

Reidy, Michael, Gary Kroll, and Erik Conway. *Exploration and Science: Social Impact and Interaction*. Oxford: ABC-CLIO, 2006.

The Study of Workplaces and Scientific Practice

The notion of work practice explored in this book developed from a wide variety of research on the nature of work, knowledge, and learning since the 1950s. Some years after the application of these ideas in Europe, particularly Scandinavia, these perspectives on the nature of knowledge and human performance began to influence the design of workplaces and computer tools in the United States (ordered alphabetically):

Hutchins, Edwin. *Cognition in the Wild*. Cambridge: The MIT Press, 1995.

Jordan, Brigitte. "Technology and Social Interaction: Notes on the Achievement of Authoritative Knowledge in Complex Settings." Menlo Park, CA: Institute for Research on Learning, Technical Report No. 92-0027, 1992. http://www.lifescapes.org/Writeups.htm (accessed December 2004).

Lave, Jean. *Cognition in Practice*. Cambridge: Cambridge University Press, 1988.

Luff, Paul, Jon Hindmarsh, and Christian Heath, eds. *Workplace Studies: Recovering Work Practice and Informing System Design*. Cambridge: Cambridge University Press, 2000.

Suchman, Lucy A. *Plans and Situated Actions: The Problem of Human-Machine Communication*. Cambridge: Cambridge University Press, 1987.

Wenger, Etienne. *Communities of Practice: Learning, Meaning, and Identity*. New York: Cambridge University Press, 1998.

Wynn, Eleanor. "Taking Practice Seriously." In *Design at Work: Cooperative Design of Computer Systems*, ed. Joan Greenbaum and Morten Kyng, 45–64. Hillsdale, NJ: Lawrence Erlbaum Associates, 1991.

For the specific observational methodology and analytic perspective influencing the study of the MER scientists, see the discussion and references in:

Clancey, William J. "Observations of Work Practices in Natural Settings." In *Cambridge Handbook on Expertise and Expert Performance*, ed. Anders Ericsson, Neil Charness, Paul Feltovich, and Robert Hoffman, 127–145. New York: Cambridge University Press, 2006.

Contrasting with the harmonious MER experience, this study describes the attitudes of scientists and engineers toward each other onboard an oceangoing vessel:

Bernard, H. Russell, and Peter D. Killworth. "Scientists and Crew: A Case Study in Communications at Sea." *Maritime Studies and Management* 2 (1974): 112–125.

My analysis of the Haughton–Mars Project on Devon Island is one of the few studies of a field science expedition as a work practice; it references related work. See also the study of remote robotic archeology in deep water by Mindell and Croff:

Clancey, William J. "Field Science Ethnography: Methods for Systematic Observation on an Expedition." *Field Methods* 13 (3) (August 2001): 223–243.

Mindell, David, and Katherine L. Croff. "Deep Water, Archaeology and Technology Development." *MTS Journal* 36 (3) (2002): 13–20.

The study of scientific work and the philosophy of science are very broad topics. The analytical perspective on representational mediation applied to MER is influenced especially by the work of Latour and Schön. They provide respectively sociological and learning perspectives on the nature of knowledge and representational tools. Kohler and Rudwick focus on scientific fieldwork.

Dyson, Freeman. *The Scientist as Rebel*. New York: New York Review of Books, 2006.

Kohler, Robert E. *Landscapes and Labscapes: Exploring the Lab-Field Border in Biology*. Chicago: University of Chicago Press, 2002.

Latour, Bruno. *Pandora's Hope: Essays on the Reality of Science Studies*. Cambridge, MA: Harvard University Press, 1999.

Rudwick, Martin. *The New Science of Geology*. Surrey, UK: Ashgate Variorum, 2004.

Schön, Donald. *Educating the Reflective Practitioner*. San Francisco: Jossey-Bass Publishers, 1987.

Finally, the reflective scientific work of Oliver Sacks strongly influenced my interviews of the MER scientists and engineers: to understand "the endless forms of individual adaptation by which human organisms, people, adapt and reconstruct themselves, faced with the challenges and vicissitudes of life" (xvi). The title describes how an autistic feels in trying to understand this "intricate play of motive and intention" (259):

Sacks, Oliver. *An Anthropologist on Mars: Seven Paradoxical Tales*. New York: Alfred A. Knopf, 1995.

Glossary

Activity	A group of related instrument applications (Observations) at a given surface feature, such as unstowing the IDD, taking a Hazcam photo, and applying the MI.
AEGIS	Autonomous Exploration for Gathering Increased Science; 2010 upgrade to Opportunity's software that chooses rocks to photograph based on predefined features.
APGEN	Activity Plan Generator; software tool for posting daily rover plans; incorporated in MAPGEN.
APXS	Alpha Particle X-Ray Spectrometer; determines elemental chemistry of rocks and soils by direct contact with the surface.
AI	Artificial intelligence; a research area in computer science aiming to develop computer programs that replicate intelligent human behavior.
Athena	Officially, the project name for the scientific instruments carried by MER; also, the Cornell-based team that proposed the mission.
AutoNav	Autonomous Navigation; software enabling MER to visually detect and avoid hazards when driving to a specified target.
Autonomy	In robotics, the degree to which the system can carry out assigned tasks independently without detailed programming and/or supervision.
Cassini-Huygens	First spacecraft to orbit Saturn (July 2004), with programmable instruments used to study the planet, its rings, and icy moons; carried Huygens, which landed on Titan (January 2005).
Constraint Editor	Software interface for adding and editing rover plan constraints, particularly important for sol-specific requirements (e.g., completing a drive by a certain time).
CRISM	Compact Reconnaissance Imaging Spectrometer for Mars; a visible-infrared spectrometer onboard the Mars Reconnaissance Orbiter for finding minerals and chemicals that may have interacted with water in the past.
DIMES	Descent Image Motion Estimation System; used to estimate horizontal velocity from three images taken 3.5 seconds apart starting at a target altitude of 2000 m before landing.
Downlink	Process of transmitting data from spacecraft "down" to Earth.
Feature	Surface object of interest, such as rock, soil region, hill.
EOWG	Experiment Operations Working Group; meeting between scientists and engineers during the Pathfinder mission to plan a sequence for the next sol.
FIDO	Field Integrated Design and Operations; the ORTs and specifically the rover mockup used by the Athena team to rehearse science operations in investigating and driving through an unknown place.

Field science	An exploratory but systematic scientific investigation of some place (the "field"), focusing on the land, climate, artifacts, and so on in that area, according to disciplinary interests.
Galileo	First spacecraft to orbit Jupiter (1995–2003), included a probe launched into the planet's atmosphere.
Hazcam	Hazard Avoidance Camera; body-mounted, front- and rear-facing set of stereo pairs, each with a 124-degree (fish-eye) field of view located under the rover's deck, looking over the wheels. Provides terrain context for onboard navigation (AutoNav) and for programming IDD operations.
HCC	Human-centered computing; a methodology for designing work systems holistically, starting with the capabilities, methods, and interests of the people, rather than the functions and capabilities of automation.
HGA	High-gain antenna; steerable direct-to-Earth antenna on Mars enabling high-rate transmissions; compared to the omnidirectional low-gain antenna for transmitting via the Deep Space Network or to Mars orbiters (Odyssey and Mars Global Surveyor).
HiRISE	High Resolution Imaging Science Experiment camera on the Mars Reconnaissance Orbiter. Provides telescopic images of objects less a meter wide (resolution up to 0.3 m/pixel [1 ft]).
IDD	Instrument Deployment Device; the robotic arm on which are mounted the RAT, MI, APXS, and MB.
JPL	Jet Propulsion Laboratory in Pasadena, California, a NASA Center that is part of the California Institute of Technology (Caltech).
LTP	Long-term planning group, companion to the STGs, which generally planned daily sol Activities for two weeks and longer-term campaigns.
Maestro	SAP successor including geographic information system for organizing images by location and the spacecraft activity plan; includes capability for distributed planning.
MAPGEN	Mixed Initiative Activity Plan Generation; software tool used by engineers to build daily rover activity plans; incorporates the EUROPA constraint-based planning and inference system with APGEN. Enables engineers to manage and verify the multidimensional requirements in a daily science plan (especially instrument-specific flight rules and cumulative time, memory, and power).
Marsokhod	Russian-built six-wheeled rover, used by McDonnell Douglass and NASA Ames researchers in the 1990s for experimenting with rover control programs and methods for remote operations.
MB	Mössbauer Spectrometer (often simply "Mössbauer").
MER	Mars Exploration Rover, the mission name, also referring to one of the twin rovers, Spirit and Opportunity; the first mobile laboratory enabling an overland scientific expedition on another planet.

MER Exploration System	Also "mission system," includes all aspects of the mission pertaining to what will be done and how, including scientific objectives, rover and instrument design, organization, processes, schedules, software, network communications, outreach, training, testing, operations during flight, landing, and surface, etc.
MGS	Mars Global Surveyor, satellite orbiting Mars with camera, laser altimeter, TES, and magnetometer (active 1997–2006).
MI	Microscopic Imager, combination of a microscope and a camera, located on the IDD.
Mini-TES (MTES)	Miniature Thermal Emission Spectrometer, detects infrared radiation to aid in identifying mineral composition from a distance, also provides temperature profiles of the atmosphere; operates through the rover mast.
MIPL	Multimission Imaging Processing Lab, JPL's image processing lab.
MOS	Mission Operations System, work system that comprises teams following procedures using GDS ("ground data system") tools.
Mössbauer	A type of spectrometer, specialized for identifying composition and abundance of minerals containing iron. Sensor head is located on IDD, which places the instrument directly against a rock or soil.
MSL	Mars Science Laboratory, a mobile field science laboratory about five times as heavy and carrying over ten times the weight in scientific instruments as MER, launched in November 2011 with arrival on Mars in August 2012.
Navcam	Navigation Camera, mast-mounted stereo black and white cameras, with 45-degree field of view. Provides terrain context for traverse planning and pointing the Pancam and Mini-TES.
Observation	Specification in rover operations plan for applying a particular instrument to a target location with specified parameters.
ODY	Odyssey, scientific imaging spacecraft orbiting Mars, providing communications relay for MER.
ORT	Operational readiness test, role-playing exercise to test organizational processes and tools in mission operations
Pancam	Panoramic Camera, a mast-mounted high-resolution color stereo camera.
Payload	The suite of instruments carried by the rover, called the Athena Payload on MER.
PDL	Payload Downlink Lead, scientist who monitors instrument status and data received.
PDR	Preliminary design review, a technical review of progress in the development of the MOS.
Phoenix	Stationary programmable laboratory ("lander") that operated in northern Mars arctic late May through early November 2008 to confirm presence of subsurface ice and examine conditions for habitability.

PUL	Payload Uplink Lead, engineer who prepares daily commands for a particular instrument.
RAT	Rock Abrasion Tool, uses a grinding wheel to abrade weathered rock (5 cm wide, 5 mm deep) and brush away dust.
RSVP	Rover Sequencing and Visualization Program, converts instrument and target sequences into specific movements and orientations of the arm and instruments.
Robot	Machine controlled by a computer program.
SAP	Science Activity Planner, computer program used by scientists and engineers for creating Observations and Activities referring to features and targets.
Science assessment meeting	Daily meeting of the scientists to assess available data from rover operations on the current sol, describe STG interests, and review long-term plans.
Sequence	A list of computer-controlled operations for an automated system. A MER sequence is an ordered series of Activities, specified abstractly by scientists in terms of operations and parameters (e.g., timing, location, resolution, filters) and then converted by engineers to an executable program.
Sol	A martian solar day: 24 hours, 39 minutes and 35 seconds. Sol N indicates the number of sols, N, since the landing of a rover; thus sol 1 for Opportunity was sol 22 for Spirit.
SOWG	Science Operations Working Group, the daily meeting of scientists and engineers to develop the plan for the next sol, integrating input from the STGs.
STG	Science Theme Group, scientists affiliated by phenomena of interest: Mineralogy and Geochemistry, Soils and Rocks, Geology, Atmosphere.
Surrogate	Substitute or stand-in; MER is a physical surrogate (but not a cognitive surrogate) for the scientists.
TAP	Tactical Activity Planner, engineer who formalizes an integrated plan for next sol, following SOWG decisions.
Target	Specific location within a surface Feature, with coordinates used to parameterize Activities.
Tau	Measure of atmospheric opacity, hindering sunlight transmission.
Teleoperation	Controlling a device from a distance; often by direct "joysticking," but more generally includes programmed control whose purposes and methods are adapted through experience during mission operations.
Telepresence	The imagined experience of being at a distant location; for MER, projecting oneself into the rover's current state, imagining future actions, and creating plans (programs) that carry out the desired operations.
Telerobot	Teleoperated robot, a robot whose programs are transmitted and/or invoked by people from a distance.

Telescience	Scientific inquiry in which instruments are teleoperated adaptively as the investigation proceeds, adjusting for changing information, interests, data gathering opportunities, programming methods, laboratory capabilities, and remote operating conditions.
Tosol	This solar day on Mars (the equivalent of "today").
Uplink	Process of transmitting commands and/or data "up" to spacecraft from Earth.
Viz	Visualization ("virtual reality") software tool that produces 3-D views from two-dimensional stereo images; can predict shadows and simulate photographs from different angles taken by a simulated rover. Derived from MarsMap used in Pathfinder and developed into Mercator for Phoenix.
VTT	Visual Target Tracking; software that enables rover to self-correct orientation during a drive to a specified goal location by taking and analyzing images at spaced intervals.

Index

Note: f after a page number indicates a figure; t after a page number indicates a table.

facilities (*see* JPL)

"follow the water" strategy, 14, 156, 229

image statistics, 23t

lessons learned (*see* Planetary science, lessons from MER)

meetings (*see* End-of-sol meeting; Science assessment meeting; SOWG)

naming problem, 164–166

nominal, 2

operations training (*see* ORT)

organization, ix, xi, xiii, 63, 72–75, 78, 145, 176, 183, 225, 227, 229, 236, 260, 289, 295

pace of investigation, 6, 129–137, 143, 156, 157, 202, 222, 223, 240–242

Pathfinder influences (*see* Pathfinder mission)

PDR (preliminary design review), 55, 60, 65, 67, 163, 185, 266, 268, 295

press release, x, 7, 73, 77, 213, 285

principal investigator, x, 74, 77, 78

project scientist, 75, 77, 78, 243

public outreach, 206–210

science team (*see* Athena; MER scientists)

science theme groups (*see* STG)

software for surface operations, ix, 50, 54–56, 60, 61, 69, 79, 86, 115, 165, 180, 230 (*see also* Virtual reality)

staffing changes, 72, 73, 240–241

statistics, ix, 21–24

success criteria, 136, 152, 184, 185

surface operations, 24f, 31, 55, 58, 60–62, 176

textbook approach, xi, 4, 58, 68f, 142, 146–155

timeline 24f, 25

traverses, 15, 21, 22t, 36, 37, 40, 64, 65, 131, 132, 156, 159, 161, 227, 231, 244, 248

voyage perspective (see Exploration, voyage of discovery metaphor)

MER scientists

aesthetic experience, 210–215

backgrounds, 82–89

blending of identities, x, 72, 76, 81–87, 89, 96, 175, 193, 199

career perspectives, 92–93

collaboration with engineers, 163–166, 168, 182–184, 236

communal aspect of team, 4, 8, 55, 135, 142–146, 158, 166–170, 199, 223

embodiment in rover, 7, 60, 103–114, 117, 121, 232, 285

emotional experiences, 9, 50, 88, 124, 128, 171, 213–218, 252, 253, 256

experience "working with a rover," 4, 6, 94, 127,166, 195, 223

fatigue, 137, 138, 143, 156

field vs. lab perspectives, 156–161

frustration with pace, 106, 129, 130–137, 157, 159, 229, 241

historical perspective, xii, 39, 50, 88, 125, 152, 157, 184, 231, 232

identity as explorers, 85–89

imagination operating rover, 8, 106, 110, 204, 215, 223, 232

interviews, 74–76, 269

language relating to rover, 100–103, 111, 124, 128, 233

learning to work together, 31, 162–163

living on Mars time, 2, 18, 55, 57, 61, 65, 67, 156, 203, 206, 210, 240–242

mission niche, 72, 93–96, 199, 201, 202, 224, 256

notion of "promised land," 137, 160, 216, 244

Pathfinder mission experience, 4, 7, 62, 75, 77, 87, 124, 183

personal projects, 129, 200–202, 224

public persona, 208–210

relation to sociotechnological context, 230–232

team organization, 72–75

MER spacecraft (rover), 20f. *See also* MER mission; Opportunity rover; Spirit rover

anthropomorphism of, 7, 59, 125, 126, 129, 232, 233, 254

cameras, 20, 115, 124 (*see also* Hazcam; Navcam; Pancam)